Think About It

Problems in basic mathematics

Think About It

Problems in basic mathematics

Book 4

J. Hulbert and W. Wiles

 Heinemann Educational Books

Heinemann Educational Books
22 Bedford Square, London WC1B 3HH

LONDON EDINBURGH MELBOURNE AUCKLAND HONG KONG
SINGAPORE KUALA LUMPUR NEW DELHI IBADAN NAIROBI
JOHANNESBURG EXETER (NH) KINGSTON PORT OF SPAIN

First published 1983
© J. Hulbert and W. Wiles 1983

ISBN 0 435 50423 1

British Cataloguing in Publication Data

Hulbert, J.
    Think about it.
    Book 4
    1. Mathematics—1961-
    I. Title   II. Wiles, W.
    510   QA37.2

Also by J. Hulbert and W. Wiles
Check it Again
Basic Mechanical Mathematics (*with A. J. Raven*)

Typeset in Great Britain by Castlefield Press, Northampton
Printed and bound in Great Britain by Richard Clay (The
Chaucer Press) Ltd, Bungay, Suffolk

Notes for pupils

Nearly all boys and girls can work out sums when there are no words with them. Yet these **problems** (sums with words), as we call them, are just as easy.

Here are some ideas to help you tackle each problem:

1 Always read each problem carefully all the way through.

2 Each problem is a story which contains numbers. Imagine you are the person in each problem and ask yourself what you must find out.

3 If it helps you make a small drawing and try to estimate (make a guess) roughly what the answer could be.

4 Set out and work the problem carefully.

5 Read the problem once more to make sure that your answer is a sensible one.

6 Remember, you may not be able to solve the whole problem at first. See whether there is any part of it that you can understand. This may help you with the rest of the problem (rather like doing a crossword puzzle, or like a detective building up clues).

Example

In a 100 m race, Carol is given a 15 m start.
How far does she have to run?

(a) Imagine you are Carol. Will you have to run farther than the other runners, or not so far?

(b) If necessary do a drawing. This will show that you have not so far to run.

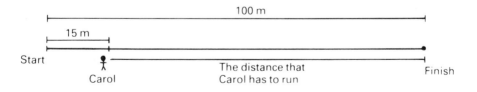

(c) To make the answer come to less than 100 m, you take the 15 m from 100 m. The answer is then 85 m.

Mathematical vocabulary for Book 4

Allotment	A small piece of land used for growing flowers and vegetables.
Appeal	Asking for money for a charity.
Arc	Part of a circle or curved line.
Ballast	Crushed rock or stone used in building a road.
Buttress	A construction of brick, stone, etc., used to support a wall.
Calculate	To work out the answer to a problem.
Callipers	A two-legged instrument for measuring diameters.
Capital	Money available for use in making more money.
Column	In this book, a vertical row of type in a newspaper or magazine.
Commission	The fee given to someone who makes a sale of goods.
Consecutive	One number or item following another without a break.
Consignment	A shipment of goods, or an amount made up to be moved from one place to another.
Continuously	Going on without stopping.
Cube	A solid shape with six surfaces, each a square.
Cubic capacity	The amount of space inside a solid shape.
Cylinder	A roller-like shape with straight sides and circular ends parallel to each other.
Deduct	To take away or subtract.
Diagonal	A straight line sloping from corner to corner.
Diameter	The measurement in a straight line across a circle, passing through the centre.
Digit	A figure. Any of the numbers from 0 to 9.
Dilute	(Of liquids) to make weaker, e.g. by adding water.
Gross	Unit of number or quantity equal to 12 dozen (144).
Ground speed	Speed over the ground for an aircraft, as distinct from speed through the air.
Hectare	Unit of area equal to 10 000 m².
Hexagon	A plane figure with six sides.
Hub	The centre of a wheel, propeller, etc.
Identical	Exactly the same.
Income	An amount of money earned in a given time.
Investment	Money deposited in the hope of it gaining interest.
Interest	Payment made for the use of borrowed money.

List price	The selling price of goods as shown, for example, in a catalogue.
Litre	Unit of capacity or volume equal to 1 000 ml or 100 cl. The volume equal to 1 kg of water.
Millilitre	Unit of capacity equal to one thousandth of a litre (0.001 litre).
Mortgage	A loan taken in order to buy a house.
Net	In this book, the amount of money left after all taxes and expenses have been paid.
Per annum	Every year, or by the year.
Perpendicular	Upright or vertical; at right angles to a given line.
Population	The number of persons living in a city, country, etc.
Prime number	A number into which only itself or one can be divided without a remainder (a number with no factors other than itself and one).
Pyramid	A construction which has a base and triangular sloping sides. A square pyramid has a square base and four sides; a triangular pyramid has a triangular base and three sides.
Radius	The distance from the centre of a circle to any point on its circumference.
Ratio	A relationship of one set of objects, numbers, etc., to another; e.g. the ratio of boys to girls in a group.
Rectangle	A plane figure with four sides and four right angles.
Rectangular	In the shape of a rectangle.
Remnant	A piece of material left over, and usually sold off cheap.
Scale	The ratio beween something real and something smaller, e.g. the distance on a map compared to the distance on the ground.
Square number	A number that is the product of a number multiplied by itself, e.g. 4, 9, 16, etc.
Survey	To plot a detailed map.
Tonne	A unit of mass (weight) equal to 1 000 kg.
Volume	The amount of space inside an enclosed shape.
Yield	The amount of a crop harvested.

Reasons for reasoning

Set 1

1 'Sebastian Coe can run one kilometre in 3 minutes, therefore he can run ten kilometres in 30 minutes.' Explain why that statement cannot be true.

2 It takes four minutes to boil an egg, therefore it takes twenty minutes to boil five eggs. Give one reason why this could be false.

3 My car travels 18 km on one litre of petrol, therefore it can travel 1800 km on 100 litres of petrol. Give a reason why this statement need not be true.

4 John weighed an apple that he took out of a bag. It weighed 324 g. There were ten apples in the bag, therefore they weighed 3 kg 24 g altogether. Give one reason why this could be false.

5 The rainfall in Kingston, Jamaica, during October was 17.5 cm, therefore the rainfall for the year was 210.0 cm. Give one reason why this is not true.

6 In the 'Grow a Sunflower' competition, Jane Ellis measured the first fortnight's growth and found it to be 3.2 cm. Therefore her plant would grow to 25.6 cm in two months. Give one reason why this need not be true.

7 Raj counted the number of words on the first page of his new paperback. The number came to 392. As there were 82 pages, he said that there were 32144 words in the whole book. Give one reason why this should not be true.

8 The weight of a bag of coins is 9.5 kg. A cashier counted the money and the total was £16.54. Another bag of coins weighed 4.75 kg and the cashier calculated that it must hold £8.27 in value. Give one reason why he could have been wrong.

9 A cargo of 96 t of oranges arrived at a port and a checker opened one of the crates which weighed 500 kg and found that 1/16 of the orange were bad. He therefore decided that 6 t out of the whole cargo were unfit to sell. Give one reason why he could be wrong.

10 A car is advertised as having an acceleration of 0 to 60 km/h in 7 s. Therefore it should reach a speed of 300 km/h in 35 s. Give one reason why this is not true.

Set 2

1 A hot water tap runs at 5 litres a minute. A bath holds 90 litres, therefore the tap can fill the bath 3 times in 54 minutes. Give one reason why this could not be true.

2   An advertisement states that you can make 3 cups of tea from one teabag, therefore 50 teabags should make 150 cups of tea. Give one reason why this statement could be untrue.

3   The stopping distance of train travelling at 50 km/h is said to be 100 m, therefore a train travelling at 100 km/h should stop within 200 m. Give one reason why this should be false.

4   There are 100 cm in one metre, therefore there are 100 cm² in one m². Say why this cannot be true.

5   A spring stretches to 22 cm when a 10 g weight is attached to it, therefore it should stretch to 110 cm when a 50 g weight is attached to it. Give one reason why this need not be true.

6   At the age of fifteen a boy's height was 1.67 m, therefore at thirty years of age it should be $1.67 \text{ m} \times 2 = 3.34$ m. Give one reason why this is false.

7   Road accidents in Great Britain during August 1980 were recorded as 30 813, therefore for the whole of 1980 the accident figures should be $30\,813 \times 12 = 369\,756$. Give one reason why this need not be true.

8   To send a parcel weighing 1 kg costs £1.10, therefore to send a parcel weighing 4 kg costs £4.40. Give one reson why this is not true.

9   The four fastest sprinters in the world have times for the 100 m of 9.8 s, 9.9 s, 9.9 s., and 10.0 s. Therefore they should hold the world relay record for 4×100 m by running $9.8 \text{ s} + 9.9 \text{ s} + 9.9 \text{ s} + 10.0 \text{ s} = 39.6$ s. Give one reason why this need not be true.

10 A student was told that one metre square is equal to one square metre. He then said that three metres square must be equal to three square metres. Show why his reasoning was false.

Metric units

Set 3

1   In a toy factory, Peter was making hexagons out of wire for a new toy. The hexagons had sides measuring 3.7 cm. How many hexagons could he make from a coil of wire 7.548 m long?

2   In our local newspaper, every six lines of type takes up 16 mm of space. The columns are 48 cm long. What is the maximum number of lines that there could be in one column?

3   In a warehouse, square boxes of sugar are stacked in rows 17.92 m long. Each box contains 7 kg of sugar and is 28 cm long. How many kg of sugar are there in 3 rows of boxes?

4   7.02 m of wood was cut into pieces 27 cm long to make racks for use in kitchens. 7 mm was trimmed from each piece during the finishing process. How much wood was wasted through trimming?

5   The circumference of a cycle wheel is 62.8 cm. How many complete turns will it make over a distance of 653.12 m?

6   A jam factory makes jam in vats holding 122.820 litres each. From each vat they fill an equal number of large jars and small jars. The large jars hold 285 ml and the small jars hold 160 ml.
How many jars of each size can be filled from one vat?

7   A machine cut pieces of wire 32 cm long for making wire baskets. It was then found that each piece was 2 cm too long. How many more pieces could have been cut from 192 m of wire if each piece had been of the correct length?

8   A total of 145.6 t of crates standing on a dockside was being loaded onto a ship by crane at the rate of 40 per hour. Each crate weighed 455 kg.
How long did it take to load all of the crates?

9   A length of tape 177.6 cm long was cut in half. The first half was cut into pieces 24 mm long and the second half was cut into pieces 37 mm long. How many pieces were there altogether?

10   A large printing firm has two stacks of paper, one white and one blue, each weighing 13.104 t. The white paper is put into piles weighing 91 kg each and the blue paper is put into piles weighing 72 kg each.
How many more piles of blue paper than of white are there?

Circles

Set 4

1   Mrs Harris, our teacher, cut out four circles, each with a radius of 17.8 cm. She placed them in a straight line side by side. She then drew a line right through the centre of all the circles.
What was the length of the line she drew?

2   Twelve sticks of rock fit exactly into a box 45.6 cm wide. What is the diameter in mm of each stick of rock?

3   I set a pair of callipers at 138 mm when I measured the diameter of a ball. What was the radius in cm of the ball?

4   The measurement from the centre to the rim of Jan's bicycle wheel was 26.4 cm. What was the diameter of the wheel?

5   The largest clock face in the world is in Beauvais, France. It is 6.09 m across. What is its radius?

6 George cut out a circle of card with a radius of 5.5 cm. What is the area of the smallest square that the circle can be fitted into?

7 Dick placed nine 10p coins in a straight line. A 10p coin has a radius of 14 mm. What was the length in cm of the line of coins?

8 A factory worker has to cut out metal circles of radius 15 mm. How many circles can he cut from a piece of metal 2.4 m long 3 cm wide?
(N.B. some metal will be wasted.)

9 The centre circle of our football pitch was marked out. It had a diameter of 8.3 m. What was the radius of the circle?

10 Fifteen cans of fruit juice each with a radius of 6.6 cm fitted across the width of the box. What was the width of the box?

Set 5

(N.B. $\pi = 3\frac{1}{7}$.)

1 A road roller wheel has a diameter of 1.54 m. How far does it travel in one turn of the wheel?

2 The circumference of a circle marked out on our playground is 11 m. What is the distance from the outside of the circle to its centre?

3 Old Tom's farm cart needed two new spokes for one of its wooden wheels. The circumference of the wheel is 284.0 cm and the spokes go from the centre to the circumference. What length of wood is needed?

4 A helicopter blade is 5.3 m long. What is the distance travelled by the tip of the blade in one complete rotation?

5 The distance from the bottom of the figure six to the top of the figure twelve on my watch is exactly 30 mm. The centre-second hand is just long enough to touch the top of the twelve. How long is the hand?

6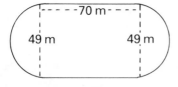

The Achilles Athletic Club marked out a running track as shown in the diagram. What is the distance around the track?

7 The hoops used in a school have a diameter of 70 cm. How many complete turns would each hoop make when bowled in a 100 m race?

8 The largest Ferris Wheel now working is in Vienna, Austria. It has a diameter of 60 m. How far would a passenger travel in seven turns of the wheel?

9

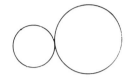

Two cog wheels, one with a circumference of 77 cm and the other 44 cm are linked in a machine. What is the distance beween the centres of the two wheels?

10 The diameter of a circular coffee table is 630 mm. Mr Gray put a metal strip all around the edge. What length of metal did he use?

Set 6

($\pi = 3^1/_7$.)

1 The circular sanding discs for my power tool have a radius of 14 cm. What is the area of each side?

2 When marking out the centre circle of a netball pitch, the contractor measured the diameter as 7 m. What area of the playground did the circle enclose?

3 When planning his garden, my neighbour cut from her lawn a circular rose bed which measured $3^1/_2$ m across the widest part. What area of garden did she dig for the rose bed?

4 The turntable in a railway yard has a radius of 10.5 m. What area does the turntable cover?

5 A machine cuts self-adhesive paper circles of diameter 7 mm. What area of material would be used for 20 of these circles?

6 A contractor has the job of providing plate glass for 32 portholes on a ship. The radius of each porthole is 28 cm. What area of glass will the contractor use?

7 A factory has to make the tops and bottoms of tins, each with a diameter of 17.5 cm. It cuts 28 circles from a sheet of metal one metre square. What area of metal sheet is wasted?

8 From a sheet of coloured card a teacher cuts 27 circles each with a radius of 5.25 cm. The card measures 80 cm by 50 cm. How much is left after the circles have been cut?

9 The clock on the town hall has three faces, each with a minute hand 1.75 m long which reaches the edge of the clock face. What is the total surface area of the three clock faces?

10 A gardener plans to have a circular lily pond surrounded by a concrete path. The path is 70 cm wide and the pond has a diameter of 4.2 m. What will be the area of the path?

Triangles

Set 7

1   The flag on David's dinghy is in the shape of a triangle. It is 24 cm long from the point of the triangle to the base. The base is 16 cm long. What is the area of the material in the flag?

2   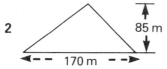 The field behind my house is shaped like this diagram. What is the area of the field in m²?

3   John was asked to cut a right-angled triangle from a sheet of cardboard. It had to be 15 cm along one side and 9 cm high. What area of cardboard did he cut out?

4   We tiled the path outside our back door with triangular-shaped tiles. Each one was ½ m along its base and its perpendicular height was the same. What area did each tile cover?

5   A blacksmith cuts a triangular piece of metal from a sheet 2.4 m long and 1.5 m wide. The triangle is 75 cm along its base and the perpendicular height is 90 cm. What is the area of the metal left in the sheet?

6   The end of a Toblerone chocolate pack is in the shape of a triangle, with a base of 3.8 cm. The height of the pack is 3.4 m. What is the area of the end of the pack?

7   James bought a kit to make a kite which came in the shape of two triangles. Each of the triangles had a base of 1.2 m and a height of 0.45 m. What was the area of the kite?

8   Mary cut from card the net of a regular triangular pyramid whose four faces were identical. The sides of each triangle measured 15 cm and the perpendicular height was 13 cm. What area of card did Mary use for her pyramid?

9   From a piece of card Helen cut a triangle which had a base of 14.6 cm. The base was twice the height of the triangle. What area of card did Helen cut?

10 A pup tent when correctly put together has a height of 1.3 m and a width of 1.5 m.
    The sides are 2.2 m by 1.5 m. What area of material is needed for the tent, not counting the floor?

Cuboids

| Set 8 |

1 A fish tank is 60 cm long and 30 cm wide. The water in it is 15 cm deep. What is the volume, in cm³, of the water in the tank?

2 A building for housing heavy machinery is to be 5 m long, 4 m wide. The foundations are dug to a depth of 2 m for extra strength, and the space is filled with concrete. What is the volume, in m³, of the concrete used in the foundations?

3 A concrete block is 24 cm long, 12 cm wide, and 7 cm high. What is the volume of three of these blocks?

4 A wooden cube has sides 25 mm long. What is the volume of wood, in mm³, used to make the cube?

5 A wooden box 28 cm long, 15 cm wide, and 9 cm high (inside measurements), is made to contain a special powder. How many cm³ of powder would be needed to fill the box?

6 A block of wood measures 15 cm by 20 cm by 8 cm. What volume of wood has a carpenter cut from the block if the volume of wood left is 2 000 cm³?

7 A trench 18 m long, 2 m wide, and 4 m deep is lined with plastic and filled to the top with water from a tank holding 200 m³. What volume, in m³, is left in the tank?

8 A cube of balsa wood has a volume of 125 cm³. What is the length of one of its sides?

9 A small tea chest from Sri Lanka is 65 cm long, 25 cm wide, and 18 cm high. What volume of tea can be packed into the chest?

10 A metal bar of square section is 1 m long and 15 cm across. How many cm³ of metal are in the bar?

| Set 9 |

1 How many cubic metres of water are there in a full tank which measures 4.5 m by 3 m by 2 m?

2 Mr Johnson wanted to concrete his garage floor which is 4.5 m long and 2.5 m wide. He decided to cover it to a depth of 0.2 m. What volume of concrete did he need?

3 A printer cuts paper 0.5 m long and 0.2 m wide for a special contract. The pile of cut paper is 0.3 m thick. What volume of paper is in the pile?

4   To store grain, a merchant had a storage bin sunk into the floor of his warehouse. It was in the shape of a cube 6.2 m wide. What volume of grain was he able to store?

5   A carpenter cut a hole 3.75 cm long, 4.5 cm wide in a slab of wood. The slab was 1.6 cm thick. What volume of wood did the carpenter take out of the slab?

6   Some containers loaded at Liverpool Docks were 16.25 m long, 3.2 m high, and 5.5 m wide. What volume of goods could be packed into each container?

7   The cold water tank in our house is 0.8 m long and 0.75 m wide. The water is always 0.6 m deep. What is the volume of water in the tank?

8   A stack of aluminium sheeting is 1.5 m high. Each sheet has an area of 2.45 m². What is the volume of the aluminium in the stack?

9   A gardener bought six blocks of peat from a garden shop. Each block was 80 cm long, 4 cm wide, and 25 cm thick. What was the total volume of peat, in m³ that the gardener bought?

10   A bin of flour in a bakery is 125 cm deep and measures 1.5 m by 1 m at the top. When the bin is half full, what volume of flour, in m³, is in the bin?

Time, distance, and speed

Set 10

(**Reminder:** distance ÷ time = speed; distance ÷ speed = time; time × speed = distance.)

1   A long-distance lorry driver averaged 86 km/h over a total driving time of 9 hours. How far had he travelled?

2   A car salesman completed the 385 km from London to Doncaster in 5 hours. What was his average speed in km/h?

3   A racing cyclist covered a distance of 174 km from London to Brighton and back at an average speed of 29 km/h. How long did the journey take?

4   A road relay team of eight runners ran 35 km in 5 hours. What was the average speed of the eight runners?

5   The average speed for a moped travelling a distance of 315 km was 45 km/h. How long did the moped take to travel the 315 km?

6   The night train from King's Cross took 7 hours to travel the 644 km to Edinburgh. What was the average speed of the train?

7 A Mini Metro travelling at an average speed of 62 km/h took 8 hours to complete a journey from Amsterdam to Hamburg. What is the distance between these two cities?

8 On a walking holiday in the Peak District, Jack Bates covered 150 km. His actual walking time was 25 hours. What was his average walking speed in km/h?

9 A helicopter flew a distance of 750 km at an average speed of 125 km/h. How long did the journey take?

10 The long-distance bus from Rome to Munich took fifteen hours to cover the 915 km. What was the average speed of the bus?

Set 11

1 How long would it take to walk 20 km at an average speed of $3\frac{1}{5}$ km/h?

2 A Trident jet airliner travelled a distance of 4 875 km in $6\frac{1}{2}$ hours. What was its average ground speed?

3 In a powerboat race off the south coast of England, one competitor covered the 312 km in $3\frac{1}{4}$ hours. What was the average speed in km/h?

4 From its take-off point to landing a glider travelled 144 km at an average speed of 45 km/h. For how many hours was the glider in the air?

5 The world hang-gliding record was made over California at an average speed of $25\frac{1}{2}$ km/h. If it took six hours, what was the record distance covered?

6 The approximate distance from Calais to Zürich is 860 km. The journey took a long-distance lorry driver a total of $14\frac{1}{3}$ hours driving time. What was his average speed for the journey?

7 During the night section of the Round Britain Car Rally, one driver travelled at an average speed of 120 km/h for $2\frac{3}{4}$ hours. What distance did he travel in that time?

8 The distance from London to Brighton is 85 km, and it was walked in a time of $7\frac{1}{2}$ hours a few years ago. What was the average speed of the walker?

9 Our journey when going on holiday last August took us $2\frac{1}{4}$ hours, and we travelled at an average speed of 76 km/h. How far did we travel?

10 On a trip of 1 440 km, our plane maintained an average speed of 432 km/h. How long was the trip from take-off to landing?

Areas

Set 12

1   When wallpapering a wall of a room 3.43 m long, a decorator used paper 49 cm wide. He managed to get three lengths from each roll.
How many rolls of paper did he have to buy, and what was the cost at £4.78 per roll?

2   The entrance hall to a house is 1.3 m wide and 5.5 m long. The chosen carpet for the hall costs £8.20 per m² plus £6.50 for fitting. What would be the total cost of having this carpet fitted to the entrance hall?

3   Vinyl tiles are sold in boxes of 20 for £1.50. Our kitchen is 4.4 m long and 2.5 m wide. Each tile is 25 cm square. How many boxes of tiles must we buy and what would it cost if an extra £7.30 is spent on adhesive?

4   A man wanting to buy a rectangular piece of land advertised as being 100 m × 75 m, found when he measured it that the measurements were actually 10% smaller than advertised. The land was for sale at 50p a square metre. How much had he saved by finding the correct size?

5   The cost of covering a road with tarmac is £5.25 per m². What does it cost to complete a road 1.5 km long and 50 m wide?

6   Sheets of veneered oak for wall covering are 30 cm wide and are sold at £1.12 for a metre length. What would it cost to cover one wall 5.4 m long and 2 m high?

7   Felt for roofing is sold in complete rolls 5 m long, 1.4 m wide. A double garage is 10 m long, 4.2 m wide. What would be the cost of felting the roof at £4.60 per roll of felt, plus £5.88 for sealing compound?

8   Timber for shelving 80 cm wide cost a carpenter £2.18 per m. He put up two shelves each 2.5 m long and 40 cm wide. What did he pay for the timber?

9   Graham bought a carpet square 3 m wide for his sitting room which is 5.5 m long and 4 m wide. To cover the surround, he bought lino paint at a closing-down sale for £2.55 a litre. If a litre covers 2.6 m², how many tins of paint did he buy and what was the total cost?

10   A school hall is 28 m long, 22 m wide, and 4 m high. Paint for redecorating costs the contractor £3.50 a litre, and a litre should cover 16 m². What would it cost for the paint for all four walls of the hall if 32 m² is subtracted for the doors and windows?

Set 13

(N.B. It might help to make sketches of some of these problems.)

1 A curtain material costs £4.50 per m². A hall has four windows each 1.4 m long, 2.6 m wide. Each window has two curtains. Allow enough material to cover each window, plus 2 m² for each curtain and calculate the cost of providing curtains for all four windows.

2 To cover a patio 11 m long and 5 m wide I bought paving stones 200 mm square. They were sold at £2.65 for 25. What did it cost me to pave the patio?

3 An alloy used in making refrigerators is sold in sheets of one square metre for £3.15. A manufacturer needed 1 000 squares each 20 cm long.
How many sheets did he need to buy and what was the cost?

4 A rectangular lawn 17 m long has an area of 98.6 m². A metal edging to surround the lawn is sold at 75p per metre. What will be the cost of completely surrounding the lawn?

5 To make business introduction cards measuring 2.5 cm by 6 cm, heavy card 60 cm long and 15 cm wide was bought at £1.35 a sheet. What would be the cost of enough heavy card to make 360 introduction cards?

6 When applying gold leaf to rare books, a bookbinder needed a sheet 2.9 cm by 1.9 cm which cost 7p per mm². What would be the cost of the whole sheet?

7 A lace border 20 cm wide is sewn onto a square tablecloth making a total area of one square metre. The lace cost 35p per 100 cm².
How much was paid for the lace border?

8 A crate was made for packing cycles for export. It was 2.5 m long, 1.4 m wide, and 0.9 m high. If the timber cost £1.50 per m², what would be the cost of the crate, allowing £1.50 for miscellaneous extras?

9 A manufacturer making cardboard boxes for items of jewellery bought card at 60p per m². The boxes were 9 cm long, 5 cm wide, and 1.5 cm deep.
How much would it cost for the card needed to make 1 000 boxes?

10 Two plate glass windows were accidentally smashed and had to be replaced. One window was 4.8 m long and 2 m high, while the other was half as long and half as wide.
What would be the cost of glass for both windows at £8.85 per m²?

Cylinders

Set 14

(Use $\pi = 3\frac{1}{7}$.)

1 An oil drum, 60 cm high, has a radius of 21 cm. What volume of oil, in cm³, can it hold?

2   One length of drain pipe 1.4 m long has a radius of 3.5 cm. What is its capacity in cm³?

3   How many cm³ of water can be contained in a cylindrical tank whose diameter is 28 cm and depth 52 cm?

4   A full jam tin contains 2 772 cm³ of jam. The radius of the tin is 7 cm. How tall is the tin?

5   A saucepan which is 15 cm deep, with a radius of 84 mm, is half full of water. What volume of water, in cm³, is in the saucepan?

6   Each cylinder in a 4-cylinder car has a radius of 3.5 cm and a depth of 12 cm. What is the cubic capacity of all four cylinders?

7   A small circular paddling pool in a garden has a diameter of 2.8 m and is 40 cm deep. The water in it is 35 cm deep.
    What is the volume, in m³, of the water?

8   Mr Graham needs a new cylindrical water tank. It must be no bigger than 42 cm in diameter, to fit into his cupboard, but he wants it to hold 76 230 cm³ of water. What is the minimum height that the tank must be in order to hold this quantity?

9   On display in a Californian museum is a portion of a giant redwood tree. It is 2.1 m in diameter and is 3.6 m long. What is the volume, in m³, of the wood in that piece of tree?

10 Before a rainstorm there was 13.8 cm of water in a water butt. After the storm there was 19.3 cm. The diameter of the butt is 56 cm.
    How many extra cm³ of water had been added by the storm?

Set 15

(Use $\pi = 3.14$.)

1   Two oil drums each have a radius of 15 cm, but one is 45 cm high and the other is 50 cm high. How much more oil, in cm³, can the larger drum hold than the smaller?

2   A tank 1 m deep, with a diameter of 52 cm, was filled with water. From this was filled a metal drum 35 cm deep, with a diameter of 22 cm. What volume of water, in cm³, was left in the tank?

3   A jug with a diameter of 6 cm and a height of 18 cm is one quarter full of milk. How many cm³ of milk are in the jug?

4   At an oil refinery the containers of crude oil are 15.8 m high with a diameter of 26 m. How many m³ of oil can each contain when full?

5   A well is 20 m deep and 1 m in diameter. It is one quarter full of water. 1.36 m³ is taken from it. What volume of water, in m³, remains in the well?

6   A grain silo is 16.2 m high. Its diameter is 12 m. It contains 120.742 m³ of grain. How many more m³ are needed to fill it?

7   A vet's syringe full of vaccine is 16 mm in diameter and is 14 cm long. An equal amount of vaccine must be used on each of four animals. How many mm³ is that for each animal?

8   A watering can 25 cm high with a diameter of 16 cm is filled from a water butt 90 cm high with a diameter of 56 cm. After the can has been completely filled 10 times, what volume of water, in cm³, will remain in the butt?

9   Andrew's mug is 10 cm tall, with a diameter of 6 cm. Sandra's mug is 11 cm tall, with a diameter of 7 cm. How many more mm³ can Sandra's mug hold than Andrew's?

10   From a tank holding 5 000 cm³ of water, a can with a diameter of 16 cm and a height of 24 cm is filled. How many cm³ of water remain in the tank?

Fractions

Set 16

1   What is the area, in km², of a field that is ¾ km long, ⁵/₁₂ km wide?

2   Mary had to multiply ¹¹/₂₄ by ⁹/₂₂. She gave the answer as ⁵/₁₆. Show as a fraction in lowest terms the amount of her error.

3   Susie and her young brother were given some sweets. Susie kept half for herself and her brother had three quarters of what was left. What fraction of the sweets did he have?

4   Matthew took ⁶/₁₅ of an hour to sharpen a saw. His supervisor said, 'You take too long. I can do that in half the time.' What length of time, given as a fraction of an hour, would it take the superviser?

5   Nita Stevens shared a sum of money among her four children. To her daughter she gave ¹/₁₆, and to each of her three sons she gave ⅓ of the remainder. What fraction of the original sum did each boy have?

6   A man set off on a 3-day walk. On the first day he walked ¼ of the total distance and on the second day he walked ⅖ of the remaining distance. What fraction of the total distance did he walk on the second day?

7   Maria and Emma were each knitting a college scarf. By the end of one week Maria had knitted ¹⁴/₂₅ of a metre, but Emma had knitted only ¹⁰/₂₁ of the amount knitted by Maria. What fraction of a metre had Emma knitted?

8   Farmer Platt had a square field with sides of 1 km. He fenced off a rectangular piece ¹⁵/₂₄ km long by ¹⁸/₃₅ km wide for his daughter's pony. What fraction of his field had he fenced off?

9 A jeweller ordered a sheet of glass to replace his broken shop window. The window was $2^5/_8$ m long, $1^1/_3$ m high. What area of glass did he need?

10 What area of wood is needed for floorboards for a room measuring $2^{11}/_{12}$ m by $2^6/_{15}$ m?

Set 17

1 A tank, $^3/_4$ m long, $^7/_{15}$ m wide, and $^8/_{21}$ m high is full of oil. What volume of oil, in m³, is in the tank?

2 Jane has a jug of milk. From it she pours $^3/_4$ of the milk. Out of this $^3/_4$ she uses $^8/_9$ to make a Yorkshire pudding. What fraction of the whole jug of milk did she use for her pudding?

3 From a wooden plank, $^7/_{11}$ is cut away. From the remainder, $^{33}/_{40}$ is cut to make a window sill. What fraction of the whole plank is used for the window sill?

4 Guy has to replace the triangular end of his pup tent which has split. The tent is $1^9/_{11}$ m wide and $1^7/_{15}$ m high. What area of material, in m², does he need for the end of his tent?

5 From a container of distilled water, two quantities had to be carefully measured out for an experiment. First, $^{15}/_{17}$ was poured into a test-tube, and from this amount $^{34}/_{35}$ was poured into a second test-tube. What fraction of the original quantity was poured into the second test-tube?

6 A grain storage bin $^{11}/_{16}$ m long, $^{13}/_{22}$ m high, and $^{24}/_{39}$ m wide is full of barley. What volume of barley, in m³, is in the bin?

7 Michelle worked out that from midday Monday to midday Tuesday she spent $^5/_{18}$ of the time sleeping. Of the rest of the day she spent $^9/_{26}$ doing school work. What fraction of the day was spent in school work?

8 What is the difference between the sum of $1^1/_4$ and $1^1/_2$ and the product of $1^1/_4$ and $1^1/_2$?

9 A hole, $3^3/_{14}$ m long, $2^4/_9$ m wide, and $1^{10}/_{11}$ m deep was dug for a swimming pool. If it were filled to the top, what quantity of water, in m³, would be in the pool?

10 A large melon was cut in half. Winston had $^{14}/_{19}$ of one half while Liz had $^{16}/_{21}$ of the other half. What fraction of the melon did each have?

Set 18

1 A coil of rope $11^1/_5$ m long is to be cut into seven pieces of equal length. How long is each piece of rope?

2   A jar of acid containing 28²/₇ litres is poured into bottles. Each of the six bottles holds an equal amount. What amount of acid does each bottle hold?

3   How many pieces of tape 2¹/₄ m long can be cut from a length of 13¹/₂ m?

4   A load of coke of 1 tonne in weight is packed into 25 kg bags.
    How many bags can be made from the load?

5   A hectare is a unit of square measure equal to 10 000 square metres. A farmer sold a field measuring 3¹/₅ hectares for building land and 32 houses were built on it. What is the area in square metres of each building plot?

6   A plank of wood with a cubic measurement of 14²/₃ m³ is sawn into eleven equal parts. Assuming that nothing is lost in the sawing, what is the cubic measurement of each piece of wood?

7   A sack of salt weighing 67¹/₂ kg is packed into containers each holding ⁵/₆ kg. How many packets can be filled from the sack?

8   A length of brocade is 53²/₃ m long and the factory makes cushion covers from the length, each of which requires ⁷/₉ m. How many cushion covers can be made from the length of brocade?

9   A gardener laying a path of concrete slabs finds that he needs exactly 37 to complete the job. The path is 59¹/₅ m long. What is the length of each concrete slab?

10 If a pile of 10p coins measures 12³/₅ mm in height, and there are 63 coins in the pile. What is the thickness of each 10p coin?

Set 19

1   Thirty-six electricity pylons are placed at equal distances from each other and they extend over a distance of 1²³/₄₀ km.
    By how many metres are they apart from each other?

2   The area of a rectangular lawn is 96¹/₃ m². It is 8¹/₂ m wide.
    How long is the lawn?

3   What must 8⁴/₅ be divided by to give an answer of 3³/₁₀?

4   A factory boiler used 8 ²/₅ tonnes of coal during an eight-week period.
    What was the average amount of coal used each week?

5   3³/₄ litres of a certain liquid weigh 30 kg. What is the weight of one litre of the liquid?

6   A parks gardener planted cherry trees 3³/₅ m apart in a line 198 m long.
    How many trees were in the avenue including the trees at each end?

7   A bill for £17.22 was paid for 5¹/₈ m of curtain material. What would be the price per metre?

8   The local bookseller built a shelf $2^7/_{10}$ m long to hold paperback books which average $^{15}/_{1000}$ m in thickness. How many books was the bookseller able to fit onto the shelf?

9   A stack of cardboard is $5^7/_8$ cm high and each sheet of cardboard is $^1/_{24}$ m thick. How many sheets of cardboard are there in the pile?

10 A church candle $^3/_{10}$ m long takes $7^1/_2$ hours to burn down completely. What length of candle will burn in one hour?

Scales

Set 20

1   John found in his atlas a map of Great Britain with a scale of 1 cm to 15 km. He measured the distance from Cambridge to Chester on the map and found it to be 16 cm. What is the actual distance from Cambridge to Chester (in a straight line)?

2   A pilot planning a flight from Southampton used a map with a scale of 1 cm to 40 km. She measured her flight path as 15 cm.
What was the actual distance she would have to fly?

3   By consulting his AA Handbook, which uses a scale of 6 km to 1 cm, Jane's dad estimated that his journey from Hull to York would be 54 km.
What measurement had he made in the handbook?

4   The scale on my atlas map of Central Europe is 1 cm to 50 km. I measure a straight line from Berlin to Warsaw as 11.5 cm. What is the actual distance between these two cities?

5   The representative of a firm has to travel from Newcastle-upon-Tyne to Melrose for an appointment. Using his road map with a scale of 4 km to 1 cm, he measures the road distance as 31.5 cm.
What actual distance would he have to travel?

6   Using an Ordnance Survey map with a scale of 500 m to 1 cm, a land agent measured a stream from Little Warlby to Streetcote as 15.5 cm.
What was the actual length of the stream that the agent measured?

7   A map with the scale of 25 km to 1 cm was used to measure the distance from Durham to Banbury. The distance was worked out at 312.5 km.
What was the measurement on the map?

8   The state boundary between Arizona and New Mexico on a map of the USA measures 5.25 cm. The scale is given as 120 km to 1 cm. How long is the state boundary between Arizona and New Mexico?

9   The air distance between Dunedin and Auckland in New Zealand is 1 050 km. On a map with a scale of 60 km to 1 cm, what distance would be measured?

10   The maths class made a plan of their classroom to a scale of 1.5 m to 1 cm. The length of the room is 9.75 m and the width is 6 m. What measurements should be shown on the class plans?

Set 21

1   The King's Cup Air Race is flown over a triangular course which measures 96 km by 72 km by 112.5 km. On a chart with a scale of 15 km to 1 cm, what measurements would the navigator accurately record?

2   The Texaco Road Map of England and Wales uses a scale of 16 km to 5 cm. Mr Simpson measured the distance on the map from Wakefield to Malton and found it to be 21.5 cm. He wanted to do the return journey.
What was the length of his journey from Wakefield to Malton and back?

3   On the AA Motorway Guide the M11 is shown as 2.4 cm in 1981. The scale used in the guide is 78 km to 3 cm.
What length of the M11 had been completed in 1981?

4   The distance from St Peter Port in Guernsey to Pleinmont on the other side of the island is 9.875 km. On a map with a scale of 4 cm to 5 km, what would be the exact measurement of this distance?

5   The school atlas shows the map of Australia with a scale of 2 cm to 75 km. The measurement from Melbourne to the capital, Canberra, is 8.1 cm. What is the actual distance beween these two cities?

6   The Potholing Club staying at the Brecon Youth Hostel wanted to have a day's potholing in the Brecon Beacons, and using a map with a scale of 2.5 cm to 8 km they measured the distance as 7.1 cm.
How far did they travel to their potholing site?

7   The distance for a Marathon race is 44.05 km. In working out a route for next year's race, the organiser used a map with a scale of 6.3 cm to 5 km. He had measured 50.4 cm on his map when he discovered that the race would not be long enough. How many kilometres was he short of the correct distance?

8   When planning to sail his boat through the Nord–Ostsee Kanal in Germany from the North Sea to the Baltic Sea, John Maybridge used a Shell map with a scale of 1.5 cm to 40 km. He measured the distance as 3.6 cm.
How many km did he travel through the canal?

9 In 1981 the James family used the French motorway when driving from Paris to Lyon. The motorway charge was one franc for every 7 km. The Michelin map with a scale of 2.2 cm to 60 km showed the distance as 15.4 cm. What did the James family have to pay fo using the motorway?

10 Using a map of London with a scale of 1 cm to 3 km, I had to decide whether the North Circular Road or the South Circular Road was the shorter way around the city. The northern route measured 11.9 cm and the south 11.4 cm. Which is the longer road and by how many km?

Similar triangles

Set 22

It is advisable in this set to make a scale drawing in your book to calculate the height or distance. Try to find a scale which will fit your book.

1 A ladder reaches to the top of a wall and makes an angle of 73° two metres from the bottom of the wall. How long is the ladder and how high is the wall?

2 A television engineer erected an aerial alongside a bungalow to avoid damaging the chimney stack. Five metres from the base of the aerial he anchored a support wire which reached to the top of the aerial. The angle between the wire and the ground is 64°. How high is the aerial?

3 To find the height of Britain's tallest building a surveyor measured a distance of 100 m from its base and measured the angle to the top of the building as 53°.
How high, to the nearest metre, is Britain's tallest building?

4 An oak tree at the end of the school playing field was the subject of a study by the science club. The height of the tree was found by measuring a distance of 14 m from the base and then finding the angle to the top of the tree, which came to 57°. How high was the tree?

5 The largest pyramid in the world is in Mexico. To find the area of the square base some archaeologists needed to find the length of one side, but they could not get near enough to measure it. Therefore they first measured a distance of 200 m at right angles from one corner, then found the angle between this corner and the next to be 65°.
What is (a) the length of one side, and (b) the area of this square pyramid?

6 A telephone engineer leans a ladder 1.2 m away from the base of a pole at an angle of 80° to the ground. How long is the ladder?

7 A group of geologists wanted to find the deepest part of the Grand Canyon, USA. If they could have found a spot exactly 2 km from the base, they would have measured the angle to the top as 47°. Using these measurements, what would they have found to be the depth of the Canyon (to the nearest 100 m)?

8   A flagpole in the centre of a city was measured by taking the angle at a point 10 m from the base to the top of the pole. This was found to be 59°. How high was the pole?

9   When painting a gutter of her house, Sheila found that her ladder was 1.5 m short of the gutter. She had set the ladder 2.5 m from the wall at an angle of 70°. How long a ladder did she really need to complete the job?

10 A rock climber on the North Devon cliffs had reached a point exactly halfway to the top when he was spotted by friends who were standing 30 m from the base of the cliff. They measured the angle from the ground to the climber and found it to be 39°. How much further did he have to climb?

Approximation and estimation

Set 23

Round off the figures before working out, to find the easiest possible way to reach an estimate. (Remember that an estimate is not an exact answer.)

1   A pools winner received a cheque for £18 656.15. What sum of money, to the nearest £100.00 did the winner receive?

2   A buyer for a firm looked at toys priced at £4.95 each. He wanted to order 500. To the nearest £100.00, what should he pay?

3   39 tonnes of cattle feed were entered on a bill at a cost of £2 384.15. What, to the nearest £10.00, was the cost of a tonne of cattle feed?

4   There were 71 stacks of boxes in a storeroom, with 82 boxes in each stack. What was the estimate, to the nearest 100, that the storeman gave of the total number of boxes?

5   A lorry driver had to estimate how long a journey of 488 km would take him if his lorry averaged 50 km/h. To the nearest hour, what answer should he have given?

6   A cargo ship carrying 1 855 t of iron ore was off-loaded into 38 railway trucks. What load of ore, to the nearest 10 t, was carried in each truck?

7   Three pieces of copper pipe 8⅔ m, 6⅘ m, and 11⅛ m were placed end to end. What, to the nearest metre, was the length of the three pieces?

8   An area of 18 955 hectares was shared among 42 representatives of a soap firm. What area, to the nearest 10 hectares, should each one cover?

9   An automatic powder dispenser held 3 858 g of talcum and was set to fill 28 containers with equal amounts. How much talcum, to the nearest 10 g, should be in each container?

10 A surveyor estimating the area of some building land measured the rectangle as 78.75 m long and 42.09 m wide. To the nearest 100 m², what was his estimate?

Revision

Set 24

N.B. $\pi = 3^1/_7$ throughout the following sets.

1 Padma bought a set of felt pens costing £2.35 and gave the assistant a £5.00 note. The change she was given was five 50p coins and the rest in 5p coins. How many 5p coins was she given?

2 The distance from London to Cardiff is 248 km and from Cardiff to Carlisle is 444 km. How far is it from London to Carlisle via Cardiff?

3 A box containing rice is 8 cm long, 5.5 cm high, and 3.5 cm wide. What volume of rice will the box hold?

4 $26 \times 8 = 208$. What is 260×0.8?

5 The population of London is 7 200 000 and of Paris 2 317 227. The population of Tokyo, Japan, is 2 184 672 more than the other two cities put together. What is the population of Tokyo?

6 A cargo of 675 t of sugar was loaded equally into 15-tonne trucks. How many truck loads would be needed to clear the cargo?

7 How far will a car travel in ¾ hour if it travels at a constant speed of 70 km per hour?

8 In planning an orchard, a gardener decided to allow 6.5 m² for each fruit tree. What area of orchard was needed for 48 trees?

9 A book was purchased at £1.80 and sold at a profit of 33⅓%. What was the selling price of the book?

10 Jason paid £2.34 for a bag of oranges. Each orange cost 13p. How many oranges were in the bag?

Set 25

1 Twenty-one sacks of potatoes weigh 493.5 kg. What is the weight of each sack?

2 A store had a roll of stair carpet 42 m long and 80 cm wide. Two lengths, 6 m and 15 m were cut off. What was the area of the remaining carpet?

3 The weight of a parcel is 8.25 kg. The packaging makes up ²/₁₅ of the weight. What is the weight of the goods in the parcel?

4 How many litres of petrol are needed for a journey of 970 km for a car that averages 9.7 km per litre?

5 Sheena won £100.00 in a competition. She decided to buy a dress for £24.27 and a suit for £57.85. How much money had she left from the £100.00?

6 Nine of Alison's paces measure 747 cm. If all her paces were of exactly the same length, for how many metres would 200 paces take her?

7 Washing machines sell for £160.00 cash, but if paid for in 12 monthly instalments, an extra 5% is added. David decided to pay by instalments. How much did he pay each month?

8 Five digital clocks in a jeweller's window showed these times—1340 h, 0515 h, 1515 h, 2025 h, and 1645 h. Only one of them is working, and 8 hours 12 minutes later it showed 2327 h. What was the time first shown on the clock that was working?

9 On the M1 motorway, 35-tonne trucks are used to move 5 000 tonnes of earth. How many full loads are needed and what load will the last truck carry?

10 New goalposts were bought for 23 schools in one county. The list price for each set of posts was £72.15, but the county was able to buy them for £65.20. How much was saved altogether by buying at the lower price?

Set 26

1 A load of 3 724 kg of potatoes is packed into 28 kg bags. How many bags can be made up from this load?

2 John and Alan were discussing which was the bigger, a cube with sides of 8 cm or a block measuring 11 cm by 6 cm by 7.5 cm. Which is the bigger, and by how many cm³?

3 I could pay for an electronic calculator in 11 weeks if I paid £2.25 each week. How long would it take me to pay for it if I paid £2.75 each week?

4 Leighton Rees with three darts threw double 19, treble 17, and a single 15. What was his total score with the three darts?

5 A chemist used 2 275 ml of cough mixture to fill bottles holding 35 ml each. How many bottles did he fill?

6 A yacht left Felixstowe at 0947 h on Tuesday and arrived at the Hook of Holland 19 hours 38 minutes later. At what time and on what day did it arrive?

7 A man left a total of £17 850 in his will. He left £375.00 to his favourite charity, a half of the remainder to his wife, and the rest to be shared equally among his five nearest relatives. How much did each of the relatives receive?

8 The price of a Sunday newspaper was increased by $\frac{1}{6}$. It now costs 28p. What was the cost of the paper before the increase?

9 A delivery of ties at a shop came to £24.64. Each tie cost the shopkeeper £1.12. How many ties were delivered?

10 In a box of 240 oranges, 15% were bad. How many oranges were fit to sell?

Set 27

1 31 new machines were installed in a factory. The machines cost £1 568.00 each, and the cost of installing them totalled £1 499.75. How much was this altogether?

2 A length of material measured 34 m. It was cut into pieces 74 cm long to make pillowslips. How many pillowslips were made and what length of material was left over?

3 Christopher, Simon, and Jake shared £4.00 in the ratio of 7 : 5 : 4. How much did each have?

4 A florist counting her day's takings found that she had two £10.00 notes, three £5.00 notes, sixteen £1.00 notes, twelve 50p coins, two 20p coins, and the rest in 5p coins. If the total amount came to £58.00, how many 5p coins did she have?

5 13 t of coke was loaded into two trucks, with one truck having 750 kg more than the other. What weight of coke was in each truck?

6 Two blazers were for sale in a shop: a blue one costing £22.00 with a 10% discount, and a black one costing £23.00 with a discount of 11p in the pound. Which was cheaper and by how much?

7 After laying a carpet in her sitting room, Mrs Jameson had a margin of 75 cm all round. The carpet measured 5.50 m by 3.75 m. What was the area of the room?

8 Of her daily allowance, Jenny spends $\frac{1}{2}$ on her lunch, $\frac{1}{3}$ on the bus fare, and the remaining 15p on sweets. How much is her daily allowance?

9 The Gota Canal in Sweden is 185 km long and the Corinth Canal in Greece is 6.5 km long. How many times could the Corinth Canal go into the Gota Canal and how much would be left over?

10 Michael sold his hi-fi equipment for £75.00 but he had paid 15% more than this when he bought it. How much had he paid for it?

Set 28

1   When mixing fertiliser for his lawn, Guy used 4/9 sand 1/3 peat, and 3 kg of special chemical. What weight of fertiliser did he mix altogether?

2   The youth club members wanted to paper the end wall of their club room. It was 6.5 m long and 4.25 m high. The wallpaper was 50 cm wide. What length of paper did they need for the wall?

3   A shop paid £351.00 for football shirts which cost £4.50 each. How many shirts were bought?

4   Out of the North American Great Lakes, Superior is the largest, with an area of 82 400 km². Lake Huron is next, with 59 575 km², and Lake Michigan's area is 58 000 km². How much smaller is Lake Superior than the other two put together?

5   The time in Bolivia, South America, is 4 hours behind Greenwich Mean Time. The time in Sri Lanka is 5½ hours ahead of Greenwich Mean Time. What is the time in Bolivia when it is 0216 h on Thursday in Sri Lanka?

6   Sarah had £2.45, Joan had £3.15, Gary had £4.28, and Ben had £1.00. They decided to put all their money together and then share it equally among the four of them. How much did each have?

7   Our deep freezer has inside measurements of 0.75 m high, 0.8 m wide, and 0.5 m deep. What is the cubic capacity of the deep freezer in m³?

8   My sports magazine sold 14 657 copies in October, 13 964 in November, and 17 368 in December. How many copies were sold in that three-month period?

9   A teacher, mixing paint for her class, used a special measure so that she could put 26 ml of paint and 70 ml of water into each pot. She used 0.91 litres of paint. How much water did she use?

10   There were 36 windows in a block of offices. Each window had double glazing at a cost of £268.52. There was also an extra large window that cost £340.65. How much altogether was paid for double glazing?

Set 29

1   A farmer had a piece of land 3/5 km by 15/16 km in three lots of equal size for development. What was the area of each lot?

2   A police car chased a getaway car out of London for a distance of 18 km. Its average speed was 72 km/h. For how many minutes did the chase last?

3   Jeremy cut a piece from a rectangular slab of wood measuring 36.4 cm by 21.7 cm. He cut it straight across from corner to corner. What was the area of the wood he cut off?

4   The stands in a supermarket each have 4 metal shelves 4.7 m long. In all, 300.8 m of metal shelving was used. How many stands and how many shelves are there?

5   A sheet of steel 6 m long, 4.5 m wide and 0.6 cm thick is used for building ships. What is the volume of steel in the sheet in cm³?

6   The single fare on the hovercraft from Ramsgate to Calais was £4.65. The cost of travelling for a party came to £74.40. How many people were in the party?

7   From a first school of 280 children, 10% were absent last Monday. Of the remainder, only 25% had school meals. The rest had packed lunch. How many children had packed lunch?

8   Shirley was tying a box of chocolates with ribbon for a present. The box was 19 cm long, 13 cm wide, and 4 cm deep. She wanted the ribbon to go once around the length of the box and once around the width with an extra 18 cm for the bow. How much ribbon did she need?

9   A rectangular sports field has an area of ³/₈ km². It is ³/₄ km long. How wide is it?

10   The luggage allowance for long distance air travelling is 15 kg. We had three cases which weighed 6.755 kg, 5.14 kg, and 7.895 kg. What weight were we above the allowance?

Set 30

1   On a coach trip there were 32 adults and 18 children. The fare for the adults was £4.87 and for children £2.59. How much was paid altogether?

2   Derek played in 8 matches during this year's cricket season. His scores were: 14, 31, 2, 56, 27, 0, 68, 18. What was his average score for the 8 matches?

3   John, Christine, and Mark were given £6.88, with John having three times as much as the other two put together. These two shared the rest equally. What was Mark's share?

4   Mr Nash had three bakery shops in different towns. On Saturday they sold a total of 2 214 loaves of bread in the ratio of 9 : 15 : 17.
How many loaves were sold by the shop that sold the most?

5   The sun sets at 1945 h on 1st September and rises at 0604 h on 2nd September. How many hours and minutes are there between sunset and sunrise?

6   From his savings account, Adam drew out ²/₁₅ of his total savings. He still has £212.55 left. How much money was in his account at first?

7   At an election, out of a possible 24 821 votes, Mr Boyle polled 9 311 votes, and Mr Miller polled 6 793 votes. By what majority was Mr Boyle elected and how many people did not vote?

8   From a bag of sweets, Tony had ¾ and his little sister Angela had ⁷⁄₁₂ of the remainder. What fraction of the bag of sweets did Angela have?

9   The gardener at Cheriton Hospital cut out a flower bed of diameter 8.4 m. What was the area of the flower bed?

10 A greengrocer paid £10.35 for 45 kg of apples and £16.38 for 63 kg of pears. How much per kg did the pears cost more than the apples?

Set 31

1   A field has an area of ¹⁄₂₈ km². It is ³⁄₁₄ km long. How wide is it?

2   Sharon, Clive, and Ann took turns to drive the 320 km when they went to visit friends. Sharon drove 40% of the total distance and Clive drove 25%. For how many km did Ann drive?

3   In an old house the window panes measured 21 cm by 15 cm. Each window had 16 panes and there were 10 windows in the house. All the glass had to be replaced after a fire. What was the total area of glass needed?

4   Newspapers are tied into bundles each evening using 306 m of string at a cost of £1.80. Each bundle uses 1.7 m of string. How many bundles are there each evening and what is the cost of the string for each bundle?

5   A farmer bought some bags of fertiliser at a discount price of £2.38 a bag. His bill came to £71.40. How many bags did he buy?

6   An aquarium has a fish tank measuring 2.25 m by 1.3 m by 0.8 m. When it is ¾ full, what volume of water, in m³, is in the tank?

7   A taxi and a car left King's Cross Station for the same destination 21 km away. The taxi averaged 35 km/h and the car 42 km/h. How many minutes earlier than the taxi did the car arrive?

8   A market stall holder in a seaside town sold oranges at 10 for 65p. Altogether he sold 15 000 during the season. How much money did she take for the oranges?

9   At a County Agricultural Show there were 28 859 visitors on the first day, 37 674 on the second, and 24 757 on the third. The organisers had hoped for 100 000 visitors. By how many were they short of their target?

10 A factory making kitchen equipment buys thin metal rods in 1 m lengths. From these are cut pieces 24.5 cm long for making plate racks. How many pieces can be cut from 50 one-metre lengths and how much will be wasted?

Set 32

1   Seaside rock is first made of ingredients mixed in large quantities, then is pulled into long lengths and cut into smaller pieces. A rock-maker at Winton-on-Sea mixed 66 kg of ingredients and from this he made an equal number of small sticks of rock, each weighing 165 g, and larger sticks, each weighing 275 g. How many sticks of each size did he make?

2   A box containing a pen and a pencil costs £1.02. A similar box holding a pen and three pencils costs £1.20. What is the cost of each pencil?

3   As a computer operator, Sonya's salary is £9 100.00 per annum, while her sister Hilary has a weekly wage of £82.50 as a hairdresser. How much more than her sister does Sonya earn each week?

4   A farmer harvested a barley crop on Monday, Tuesday, and Wednesday. The amount harvested on these three days was in the ratio of 4 : 7 : 13. Tuesday's yield was 2.8 t. What was the total yield for those three days?

5   A lorry travels 117 km in 2¼ hours. At what time did it deliver its load after leaving the depot at 12.55 p.m. and what was its average speed?

6   A strip of metal is 0.2 cm thick and 0.5 cm wide. What would be the volume of a 15 cm length of this strip?

7   Peter was the winner of the 400 m race at 45³⁄₅ s. The last man, Andy, took 51½ s. How many seconds faster was Peter than Andy?

8   Two circles are marked out on a school playground. One is half the area of the other. The smaller circle has a radius of 5.6 m. What is the area of the larger circle?

9   James was given ³⁄₅ of a sum of money whilst his brother Simon received ⁵⁄₆ of the remainder. What fraction of the money did Simon receive?

10 The night shift at a car factory started at 2045 h and finished at 0510 h the next day, with a total of 55 min for breaks. A man earned £7.50 an hour for each working hour on night shift. How much would he earn in a week of five nights?

Set 33

1   Using the figures 6, 8, 2, and 7, make the largest number possible, divide the result by 28, and then multiply by 5.6.

2   The minute hand of a clock travels through 30° of arc (of a circle) every five minutes. A clock is started at 0810 h. What will be the time after the minute hand has travelled through 210°?

3   What fraction of a m³ of water is contained in a full tank when the tank measures ¹¹⁄₁₅ m by ⁵⁄₁₂ m by ⁹⁄₂₂ m?

4 A factory sent out a number of identical parcels at a cost of £47.97. Each parcel cost £1.17 to send. How many parcels were posted?

5 On a train journey from Edinburgh to London there were 940 passengers when it started. 30% left the train at Newcastle-upon-Tyne, but 62 people got on. 40% of the remaining passengers left the train at Peterborough and 120 got on. How many were on the train as it continued to London?

6 A headteacher bought a cassette recorder for the school for £22.14, which was 3/16 of the money in the school bank. He spent a quarter of the remainder on books. How much did he spend on books?

7 Mr Fraser owned two furniture shops. A sales representative asked him to try some new curtain material so he bought 124 m and sold it at £3.50 per metre. The first shop sold three times as much as the second. How much money did the first shop take for the material?

8 Tom earns £300.00 every four weeks at his work. Each week his bus fares cost £3.25, his lunches cost £7.50, and he gives his mother £25.00. How much has he left over each week?

9 A square sheet of aluminium with sides of 20 cm has six circles cut from it. Each circle has a diameter of 28 cm. What area of aluminium is left?

10 Mrs Fenton keeps her house accounts very carefully. This week she spent 4/15 of it on rent and rates, 1/10 on heat and light, and 6/15 on food. She had £15.00 left for other items. How much does she have to spend each week?

Set 34

1 Peter drove 260 km on the first day of his four-day holiday. The distance driven on each of the four days was in the ratio of 4 : 1 : 7 : 3. What was the total distance that Peter drove on his holiday?

1 Mr Lock took an average amount of £155.00 per day in his shop last week, when he opened for 6 days. This week he can open for only 5 days because of a holiday. If he is to take the same total amount as last week what must be his average daily takings?

3 Mount Everest is 8 850 m high and Ben Nevis is 1 340 m high. How many times higher is Everest than Ben Nevis?

4 A gardener had two equal lengths of metal edging, each 21.4 m long. The first length was used to make a square flower bed and the second length to make a rectangular flower bed 7 m × 3.7 m. Which flower bed had the larger area and by how much?

5   The longest-ever snowmobile trip began at 0200 h on 4th December 1977 on the Pacific Coast of the USA and ended on the Atlantic Coast at 1700 h on 4th February 1978. Allowing for the fact that time on the west coast is three hours behind that on the east coast, how long did the journey take in days and hours? (N.B. Subtract the time difference.)

6   There are 256 children in our school and $^{15}/_{32}$ of them are girls. $^{5}/_{24}$ of the girls travel to school by bus. How many girls travel by bus?

7   A plastic container of washing-up liquid 15 cm tall with a diameter of 56 mm is three quarters full. What volume of liquid, in mm³ is in the container?

8   A bank manager has a pile of new five-pound notes numbered consecutively from DT664400 to DT664770. How many five-pound notes are in the pile?

9   Our first stop after leaving Calais was just north of Paris, a distance of 255 km. We did the journey in 3¾ h. What was our average speed?

10   The NO ENTRY sign at the end of out street has a diameter of 35 cm. What is the surface area of the sign?

Set 35

1   On the fourth stage of the Tour de France, the Irish cyclist, Kelly, won in a time of 6 h 20 min. His average speed was 18 km/h. What distance did he cover in the fourth stage of this cycling race?

2   Our water tank which is 1.5 m wide by 2.3 m long froze over last winter and the ice was 3 cm thick. What was the volume of the ice on the tank?

3   A prize of £540.00 was shared by the three winners. Mr Harris's share was $^{2}/_{5}$, Mr Jones's share was $^{5}/_{12}$, and the rest went to Mr Wilson. How much money did Mr Wilson receive?

4   The largest four-faced clock in the world is in Milwaukee, USA. Its minute hands are 6 m long and their tips are 30 cm from the circumference of each face. What is the total area of the four clock faces?

5   A box has a cubic capacity of $^{1}/_{3}$ m³. It is $^{5}/_{12}$ m long, $^{3}/_{10}$ m wide. How deep is it?

6   A store paid £68.16 for 32 m of cloth which was sold at £3.08 per metre. How much profit was made on each metre?

7   A handbag maker needs 80 strips of plastic each 25 cm long and 15 cm wide. He cuts these from a sheet 2 m square. What area of plastic is left?

8   A traffic survey in Stanforth showed that in one hour 90 private cars and 35 vans or lorries passed along the main street. What percentage of the total number of vehicles was made up of private cars?

9   Raj had three identical presents to tie with coloured ribbon. Each present measured 21 cm long by 15 cm wide by 8 cm deep. He wanted the ribbon to go once around the length and once around the width with 17 cm to spare for a bow. How much ribbon did he need for all three presents?

10 The cost of manufacturing a dining chair is £8.05. How many dining chairs can a factory make for £837.20?

Set 36

1   A transcontinental coach left London at 0738 h on Monday and arrived in Belgrade, Yugoslavia, 4 days 6 h 53 min later. At what day and time did it arrive (London time)?

2   During the summer season the Albany Hotel, Blackpool, uses an average of 615 eggs every day. At this rate, how many eggs would be used during the months of July and August?

3   The mother of Alan, Janet, and Clive gave them £5.00 to share among themselves, with Alan having 4 times as much as the other two together, and Janet having 3 times as much as Clive. How much more than Clive did Alan have?

4   A garage ordered a parcel of plugs which cost £2.75 a set, plus three plug spanners at 86p each. The bill came to a total of £90.58. How many sets of plugs did the garage order?

5   Hilary mixed herself a cool summer drink with 180 ml of squash, 300 ml of soda water and 420 ml of iced water. Express these amounts as a ratio.

6   Paul cut a piece from a rectangular slab of wood. He cut it straight across from corner to corner. The piece of wood was 42.5 cm long and 35.2 cm wide. What was the area of the piece he cut off?

7   Jill and her sister Lucy wanted to buy their mother a present. Jill had £1.50 and Lucy had $7/10$ of this. The present they wanted cost £3.10. How much more money did they need?

8   A pony is put out for grazing and is tethered to a post. It can graze only to the length of the rope which is 6.3 m long. What area of grazing does this allow the pony?

9   The area of the bottom of a tin is 135.6 cm². It is 16 cm deep. If the tin is $2/3$ full of oil, what is the volume of oil in the tin?

10 Julian said to his little brother Paul, 'I feel generous today. I shall give you one twentieth of one tenth of one fifth of a kilogram of my toffee.' What weight of toffee did Paul have?

Set 37

1 A water tank holds 65 litres when it is ⅙ full. How many litres of water does it hold when it is ¾ full?

2 On each of the four days of a cycling holiday, two sisters each covered 36 km, 52 km, 27 km, and 41 km. What was the average daily distance covered, and what would the average have been if they had covered the same distance in only three days?

3 Two clocks were each set to the correct time at the 7 a.m. time signal. One clock gained 7.5 s an hour and the other lost 15.5 s an hour. How far apart in time were they at 7 a.m. the next day?

4 The Aston Villa supporters' club received 746 applications to travel by coach for an away game in London. The secretary ordered 42-seater coaches for the journey. How many coaches were needed and how many spare seats were there?

5 To run a 100 m race in 10 s is the target for most good sprinters. What is the average speed in km/h for this sprint?

6 A large supermarket bought a number of LP records at £3.25 each. The total cost was £246.20 including £2.45 transport costs. How many LP records were paid for?

7 Brenda cut off ¾ of a length of ribbon, and of this she used ⁸/₁₅ for a dress. The length she used was 40 cm. How long was the original piece of ribbon?

8 A can is ¼ full of oil. The can measures 23 cm by 11.5 cm by 6.2 cm. What is the volume of oil in the can?

9 A gasometer (container for household gas) when full, is 11.2 m high, with a diameter of 7.7 m. How many m³ of gas does it hold when full ($\pi = 3\frac{1}{7}$)?

10 Cathy's car does 11.5 km to 1 litre of petrol. How much petrol will she use on a journey of 2 760 km?

Set 38

1 Out of 5 hours of work during a school day, Adrian calculated that he spent 87 minutes doing maths and English and 45 minutes doing history. What percentage of the school day was spent in other activities?

2 A garage mechanic put 44.8 litres of oil into three different sized oil drums in the ratio of 5 : 18 : 23. How much oil was in the smallest drum?

3 A Jaguar fighter crossed the North Sea in exactly fifteen minutes. The navigator worked out the distance as 237 km. What was the average speed for the flight?

4   The fish pond in Bill's garden is circular and has a diameter of 11.2 m. All around the pond is a concrete path 2.1 m wide. What is the area of the path?

5   Kevin and Bina were given some money. Kevin had ¾ of the money and Bina had half of what was left. She received £1.26. What was the original amount that they were given?

6   Sally saved 75p a week towards an electric toaster costing £26.25 that she wanted to buy for her mother's birthday, but by the time she had saved what she thought was enough, the price of the toaster had increased to £28.00. How much should she have been saving each week in order to meet the new price?

7   My car now has 6 958 km on its odometer. The garage serviced the car after the first 1 000 km and then it is due for servicing every 10 000 km. How many km should I travel before the next service?

8   One litre of water weighs one kilogram. A two-litre bottle of water is ⅖ full. What is the weight in grams of the water in the bottle?

9   A floor-layer uses oak strips 10 cm wide to cover the entrance hall of a house. The hall is 8.5 m long and 1.6 m wide. How much will it cost to cover the floor at £1.06 per metre length?

10 Two gardens each had the same perimeter, 160 m. In the first one the length was three times the width and in the second one the length was four times the width. What was the area of each garden?

Set 39

1   It costs £107.90 to buy a set of children's encyclopaedia, or, for an extra charge of £2.80, you could buy each volume separately at £6.15 each. How many volumes are there in the set?

2   Jason was weaving a special cloth on his own loom. On the five days from Monday to Friday he completed 1.23 m, 0.98 m, 1.54 m, 1.36 m, and 1.29 m. He charged £19.00 per metre. What were his average daily earnings?

3   An automatic printing machine prints 105 leaflets per minute. If it were able to operate continuously from 0815 h to 1007 h how many leaflets would it print?

4   A load of ballast for road-making should weigh 13.75 t. A contractor had 6.35 loads delivered to complete the piece of road he was working on. What was the total weight of the ballast delivered?

5   A water tank 47 cm high with a diameter of 28 cm is full of water. It is emptied into another tank of the same height but with a diameter of 35 cm. How many more cm³ of water are needed to fill the larger tank?

6   A five-litre can of oil is 9/10 full. 5/18 of this amount is used. How many ml of oil are left in the can?

7   Curtains for a motel cost £3.45 each to be made up. Each used 1.45 m of material. Altogether 63.8 m of material was used. How much did the motel owner have to pay?

8   A cube of sugar is 1.5 cm wide. What is the volume of the cube and how many cubes can be put into a box with a volume of 540 cc?

9   An air race was flown over a right-angled triangular route. The distances measured from the right-angled corner were 23.4 km and 17.8 km. What area of land was enclosed by the complete route?

10 Guy bought a square metre of thin copper sheet for £5.25. He cut off a rectangle $^{20}/_{27}$ m by $^{18}/_{25}$ m. What was the value of the piece that he cut off?

Set 40

1   A chemist mixed two chemicals with a quantity of water in the ratio of 3:4:7. The amount of water in the mixture was 123.2 ml. How many ml made up the complete mixture?

2   How many days were there altogether in the years 1976, 1977, 1978, and 1979?

3   The triangular end panel of a tent measures 1.35 m at its highest point and 1.62 m along the ground. What is the area of material needed for the panel if an extra 0.5 m² is added for sewing the edges?

4   At the church fair, ¼ of the money was taken on the stalls, ⅝ of the money on the sideshows, and the remaining £42.00 on refreshments. How much money was taken altogether?

5   The 175 electricians in a washing machine factory were each given a rise of 3p per hour. How much extra is needed to pay all of the electricians each week if they work a 35-hour week?

6   Mr Allen bought a sheet of Formica 80 cm by 60 cm to cover a circular table which is 56 cm across. What area of Formica was left over?

7   The 0800 hovercraft left Ramsgate on time and arrived fifty minutes later at Calais. This is a crossing of 24 km. What was the average speed of the hovercraft in km/h?

8   Garden fencing sold at £11.52 for panels 2 m long and 1.8 m high. How much was this per m²?

9   A teacher mixed 962.5 cm³ of paste to put into small containers 10 cm high with a radius of 17.5 mm. How many containers did she need if she filled each one completely ($\pi = 3^1/_7$)?

10 Flour is specially packed in 750 g bags for a supermarket to sell at 35p per bag. How much money would the supermarket take for flour if it sold 345 kg?

Set 41

1   The first solo flight of the Atlantic was completed by Charles Lindbergh who left the east coast of America at 12.52 p.m. on 20th May 1927 and arrived in Paris at 10.21 p.m. on the next day. How long did the flight take? (N.B. Subtract the time difference of 6 h between Paris and eastern US.)

2   A number multiplied by 22 is equal to $\frac{2}{3}$ of 990. What is the number?

3   A coke boiler burned $2\frac{2}{3}$ t of fuel in October and $1\frac{1}{2}$ times this amount in November. What was the total amount burned in these two months?

4   The evening flight from Heathrow to Frankfurt took 1 h 55 min. Its average speed was 324 km/h. What was the distance flown?

5   Richard's fish tank is a cube with a side measuring 40 cm. What would be the volume of water in the tank if it were completely full? (Answer in cm³ and in litres.)

6   A builder bought a plot of land for £1 375.00 and built on it a bungalow which cost him £14.785.00. He sold the property for £20 000.00. What profit did he make?

7   A pair of identical Mongolian stamps are each triangular in shape. The perpendicular measurement from tip to base is 31 mm and the base is 59 mm long. What area in cm² is covered by both the stamps?

8   The cost of a ticket for the Scotland and Wales Boys' International Match was £1.56. A party from our school paid £39.00 altogether. How many were in the party?

9   The Tour de France Cycling Race included one stage from Montpellier to Toulouse via St Pons. The organisers measured on a map the distance from St Pons to Toulouse as 6.3 cm, and from Montpellier to St Pons as 5.7 cm. The scale of the map was 3 cm to 50 km. What was the distance to be cycled from Montpellier to Toulouse?

10   550 people visited the shop at the local museum. 36% of them bought postcards only and 25% of the remainder bought other souvenirs. How many bought souvenirs?

Set 42

1   A box with a cubic capacity of $\frac{2}{5}$ m³ is $\frac{22}{25}$ m long, $\frac{9}{11}$ m wide. How deep is it?

2   For an experiment with plant types at an agricultural college, a square plot of land with an area of 2 500 m² was divided into strips 6.25 m wide. Each strip was divided into equal-sized plots and each plot had a different kind of plant. How many different kinds were there?

3   Jane's father won £15 000.00 on the football pools. He spent $^{17}/_{30}$ on a new car and $^4/_5$ of the rest on goods for the house. How much money had he left?

4   Six cricket bats and six cricket balls cost a total of £85.50. Each ball cost £2.65. What is the cost of each bat?

5   The area of a rectangular field is two hectares. The field is 50 m wide. What is the perimeter of the field?

6   A teacher had six sheets of card, each 949 cm². He gave an equal area of card to each of 26 children. What area of card did each have?

7   It was in 1978 that a balloon crossed the Atlantic Ocean, setting off from the USA at 1745 h on 12th August and arriving in Europe at 1051 h on 17th August. How long in days, hours and minutes did the record flight take? (N.B. Subtract the time difference of 6 h between Europe and the USA.)

8   A drum of oil has a diameter of 70 cm and is 1½ m tall. What is the total surface area of the metal used in the drum?

9   When $^1/_9$ of the paint has been used from a tin 17 cm tall with a radius of 7 cm, how many litres of paint remain in the tin?

10   At a silversmiths, a roll of sheet silver with an area of 787.5 cm² was divided among four workmen, with the senior having the largest piece and the junior the smallest. It was divided in the ratio of 17 : 25 : 38 : 45. How many more cm² did the senior have than the junior?

Set 43

1   A box, 84 cm long, 24 cm wide, and 14.4 cm high is full of cartons of macaroni. Each carton is 42 cm long and 4.8 cm square at the ends. How many cartons are in the box?

2   What number must be added to 46, 31, and 27 so that the average of the four numbers is 36?

3   Carpet is sold in widths of 75 cm. What would be the cost of covering a floor with an area of 42 m² if the price of the carpet is £2.30 per metre length?

4   There are two ferry crossings from Hirtshals in the north of Denmark across to Norway. One measures 4.5 cm to Arendal on the Shell Europe Map, and the other measures 5.1 cm to Kristiansand. The scale is given as 1.5 cm to 40 km. Which is the shorter journey and by how many km?

5   A prize of £27.00 was shared by three girls. Nita's share was $^7/_{12}$ of the total; Mary received $^4/_7$ of the amount that Nita had, and Linda received the rest. How much was Linda's share?

6   A number multiplied by 18 is equal to one quarter of 2 664. What is the number?

7   A ship's speed is measured in knots (nautical miles per hour). A cruise liner sailed at an average speed of 15 knots for 12 h 40 min. How far had it travelled in that time?

8   The manager of large store bought a total of 260 kg of sugar. Out of this there was four times as much white sugar as brown and cube put together, and three times as much brown sugar as cube. How much cube sugar was there?

9   A small furniture shop checking the sales of new material found that it sold an average of 20.35 m on each of five days last week. The sales so far this week have been 15.5 m, 21.6 m, 11.9 m, and 32.4 m. How much should it sell on the fifth day in order to equal last week's average?

10 A photographer's developing tray is 80 cm long and 20 cm wide. She pours 2 litres of water into the tray. How deep is the water?

Set 44

1   A container with a capacity of 15¼ litres is used to empty a tank containing 198.25 litres of water. How many times would the container need to be filled in order to empty the tank?

2   The sum of two numbers is 17 and their product is 72. What are the two numbers?

3   £11.30 was shared among four children in the ratio of 13 : 30 : 52 : 18. How much was received by the child with the largest share?

4   Brian had nearly paid off the mortgage on his house, and during 1981 he had paid a total of £335.00 to the building society. This was made up of 55% for repayment of capital, 35% interest charges, and 10% insurance. What were the three amounts?

5   Four triangular pieces of aluminium are cut from a rectangle with an area of 1 m². Each of the triangles has a perpendicular height of 35 cm and is 28 cm along the base. What area of aluminium is left?

6   The total of three numbers is 33 178. Two of the numbers are 4 563 and 8 694. What is the other number?

7   Tins of soft drink, 10 cm tall, with a diameter of 5.6 cm were priced at 32p. Larger tins, 20 cm tall, with a diameter of 11.2 cm were on offer at £2.50. Mary bought one large tin and Guy bought the same quantity of drink, but in the smaller tins. Who had the better buy, and by how much?

8   The weight of nuts in a bag is 1³/₁₀ kg. Charles uses ⁵/₂₆ of these when making a fruit and nut cake. How many grams of nuts are left?

9   A local organisation for helping people in need had a capital of £4 288.50 in the bank but it had to pay out £95.30 each month. For how long would the capital last if no money came into the fund?

10   A hovercraft was taken through West Africa and made a journey of over 8 000 km between 15th October 1969 and 3rd January 1970. How many days were taken for the journey?

Set 45

1   Mr Gale and his wife each inherited £2 750.00 from a relative. Mr Gale invested his share at 12% interest while Mrs Gale invested hers at only 9% interest. At the end of one year, how much more did Mr Gale have than his wife?

2   An assistant in an ironmonger's shop had three boxes of nails which had to be put in equal amounts into 12 plastic bags. In the boxes were $4\frac{5}{6}$ kg, $2\frac{1}{2}$ kg, and $1\frac{2}{3}$ kg. What fraction of a kg should each bag have contained?

3   Write the largest whole number less than 90 000 which begins and ends with the figure 4. How many times does the greater 4 contain the lesser 4?

4   A New York skyscraper is 176.3 m high. It is divided into blocks of offices each 2.15 m high. Each block contains 12 separate offices. How many offices altogether are in the building?

5   A car tester at the Jaguar factory was asked to test-drive a car for one hour at 55 km/h and for two hours at 70 km/h. What was the average speed for the test?

6   In a wood there are 17 beech trees, twice as many oak trees, and nine times as many other trees as there are beech trees. How many trees are in the wood?

7   A sheet of paper measuring $\frac{5}{8}$ m by $\frac{4}{5}$ m is cut into quarters. What is the area of each quarter in cm²?

8   A school aimed to collect £45.00 by Christmas for the Save the Children Fund. With only three weeks left they had collected £27.93. What was the average amount needed in each of those weeks if the target was to be met?

9   To get from Westminster Bridge to Marble Arch an American tourist took a taxi which charged 35p per km. On the Esso Road Map of London the scale is 1 cm to 160 m. The map distance is 19.5 cm. How much, to the nearest 1p, should the tourist have paid for the journey?

10   The area of the bottom of a tin is 95 cm² and it is 20 cm high. How many litres of water can the tin hold when it is full?

Set 46

1   The answer when a number is divided by 17 is equal to $\frac{1}{3}$ of 255. What is the number?

2   From a box of 9 kg of tomatoes, $\frac{1}{18}$ of them were bad and $\frac{3}{4}$ were sold on the first day. What weight of tomatoes was left?

3   A piece of copper pipe with a diameter of 7 cm has a cubic capacity of 557.5 cm³. How long is the pipe?

4   A smallholder started selling goats' milk at his farm gate. During the first four weeks he sold a total of 549.5 litres in the ratio of 15:37:43:62. How many litres did he sell in his best week?

5   A piece of card in the shape of a triangle has a base three times as long as its height. If the base is 16.5 cm long, what is the area of the piece of card?

6   When a school booked a theatre visit costing £1.25 per child, $\frac{3}{5}$ of the children wanted to go and they paid a total of £243.75. How many children were in the school?

7   The area of a rectangular playing field is 177 100 m². It is 550 m long. Mr Shah, the games master, runs three times around the perimeter each morning. How far does he run?

8   In one week there were 1 460 attendances at the local swimming pool. Out of these there were 284 more men than women and children together, and there were three times as many women as children. How many children were there?

9   A store receives 19 cartons of biscuits, and in each carton there are 36 packets. In each packet there are 7 cream biscuits and 7 plain ones. How many plain biscuits were in those 19 cartons?

10   The local council decided to use a rectangular piece of land, 420 m long, 310 m wide for allotments. Each allotment was to be 35 m long, 15½ m wide. How many could there be on that piece of land?

| Set 47 |

1   A high-speed drill cuts through a piece of steel 13.962 cm thick at a steady speed of 53.7 mm a minute. How long does it take to complete the task?

2   Calculate the number of minutes in one week.

3   A coach left Peterborough for London, 140 km away, and travelled at an average speed of 60 km/h. On the return journey it travelled at an average speed of 70 km/h. How much quicker was the return journey than the outgoing one?

4   A blacksmith had to put three metal bands around a cylindrical wooden cask. Each band had to have a 2 cm overlap for riveting. The diameter of the cask was 56.7 cm. What length of metal did the blacksmith use ($\pi = 3\frac{1}{7}$)?

5   A goods train travelling from Carlisle to Bristol took 2⅖ h to reach Manchester, where it stopped for ¾ h. It took another 3⅓ h to reach Shrewsbury, where it was delayed for 1½ h. The rest of the journey took 2⅘ h. How long, in hours and minutes, did it take for the whole journey?

6   100 cubes of silver with sides measuring 4 cm are melted down and poured into moulds measuring 20 cm by 5 cm by 4 cm. How many moulds of this size could be filled with the silver?

7   On a National Geographical Magazine map with a scale of 1 cm to 75 km, the three wheat provinces of Canada are found to be, from east to west, Alberta 607.5 km, Saskatchewan 540 km, and Manitoba 652.5 km. What are their measurements on the map?

8   The cost of manufacturing a cycle tyre is £4.65. How many tyres can be made for £1 000.00 and how much money would be left over?

9   The longest game of snooker was played for charity by two medical students who began playing at 1415 h on 26th March and ended at 0514 h on 1st April. How long, in days, hours, and minutes, did the game last?

10   A machine packed 110 kg of rice into packets containing 750 g each. Allowing for spillage of 800 g during the packing, how many packets of rice were made up?

Set 48

1   A sculler rowed upstream at 4 km/h and back to the starting post at 6 km/h and took 50 min altogether. How far upstream did the rower go?

2   The smaller of two numbers is 46 and their difference is 15. What is the larger number, and what is the product of the two?

3   Two remnants of carpet are offered at the same price. One is 1.5 m long with a perimeter of 5 m and the other is twice as long but half as wide, with a perimeter of 7 m. Which is the better bargain?

4   A manufacturer of plastic sacks used 85 cm lengths of plastic for each sack. He had an order for 264 sacks but he had only 152.6 m of plastic. What length did he still need?

5   A cylindrical water tank is 2.8 m in diameter and 2 m deep. How many m³ of water are in the tank when it is 0.5 m from the top $(\pi = 3\frac{1}{7})$?

6   In one month, five agents for an insurance company earned a total of £7 920.00 in commission, in the ratio of 62:13:24:36. What was the difference between the largest and the smallest amounts?

7   Ann invested £1 500.00 in her bank to gain interest of 11% in one year. At the end of the year she added her interest to the £1 500.00 and left it in the bank. How much money did she have at the end of the second year?

8 The average annual earnings of six company employees was £7 757.50. The first five earned £10 142.00, £9 564.00, £8 008.00, £7 150.00, and £6 275.00. How much did the sixth employee earn?

9 A glazier uses $7/16$ kg of putty on each of 8 windows, and $5/8$ kg on each of 9 larger windows. What weight of putty, in kg and g does the glazier use altogether?

10 The sides of a rectangular lawn are in the proportion of 3 : 4. The lawn has a perimeter of 22.4 m. What would be the cost of turfing the lawn at £2.00 per m²?

Set 49

1 Mr Kelly's daughter wanted a paddock for her pony, so he divided a rectangular field with a fence from corner to corner. The field was 136.4 m long and 127.8 m wide. What was the area, in m², of the paddock?

2 At the first home match, $13/18$ of the spectators were adults. Out of the children, there were four times as many boys as girls and the number of boys was 636. What was the total number of spectators at the match?

3 The sale price of a suite of furniture is 25% less than the ordinary price. The sale price is £327.00. What was the ordinary price?

4 Our church clock strikes the hours only. How many strokes of the bell will be heard on any one day of the week?

5 King Hussein of Jordan was born on 14th November 1935 and came to the throne on 11th August 1952. What was his age in years, months, and days when he became king?

6 The bill for four people at a restaurant came to £28.00, plus 15% VAT, plus 10% service charge. What was the final amount?

7 Two friends argued about which desert was the farthest across; the Sahara or the Australian. On a map with a scale of 1 cm to 40 km the Sahara measured 7.4 cm and on a map with a scale 1 cm to 30 km the Australian measured 11.7 cm. The friend who chose the Australian desert won, but by how many km?

8 A bedroom has a volume of air of 52.5 m³. The room is 6 m long and 3.5 m wide. How high are the walls of the room?

9 By how many is 8 000 greater than nineteen times 234?

10 A tin 12 cm tall with a diameter of 42 mm containing fruit salad costs 57p. A smaller tin with the same diameter but only 8 cm tall costs 38p. Is the larger tin a better buy than the smaller tin?

Set 50

1 Five children each had to complete a test and they took the following lengths of time to complete it—$3/4$ h, $1/2$ h, $5/8$ h, $7/12$ h, $5/6$ h. What was the average time taken, in fractions of an hour, to complete the test?

2 Mr Hansen makes children's swings that use four metal tubes each 3.7 m long and one tube 2.6 m long. He had 504.6 m of tubing. How many complete swings was he able to make?

3 Add 25 gross to 116 score and find $5/8$ of the result.

4 The scout group started a sponsored cycle ride at 0842 h and covered 34.68 km at an average speed of 10.2 km/h. They had two 10-minute breaks. At what time did they complete the ride?

5 A metal wheel with a circumference of 88 cm has a hub with a diameter of 3 cm. What length of steel wire would be needed to make 72 spokes reaching from hub to rim?

6 In a scientific experiment it was discovered that a squash ball bounces to $2/3$ the height of its fall. What should be the height of its second bounce after it has been dropped from a height of 1.98 m?

7 Lawn fertiliser is applied to lawns at the rate of 80 g per m². How many 1 kg bags of fertiliser are needed for a lawn 25 m long and 17 m wide?

8 What is the difference between 256^2 and 256^3?

9 The inside circumference of a wooden ring is 35.2 cm and the outside circumference is 39.6 cm. What is the thickness of the ring?

10 The cost of two dozen new textbooks is £46.80. The headteacher ordered 33 new books. How much did the bill come to?

Set 51

1 Louise had two stamp albums, one with 6 439 stamps and the other with 2 765 stamps. She sold 1 861 of her stamps and re-arranged the remainder with $4/7$ in one album and $3/7$ in the other. How many were in each album?

2 Mr Samsom's garden is $12^4/25$ m long, $7^1/2$ m wide. One third of it is used as a vegetable plot. What is the area of the remainder?

3 The cost of sending goods by rail is worked out by tonnage and distance. If 27 t of steel can be carried 300 km for £81.00, how much should it cost to carry 45 t for a distance of 500 km?

4 A saloon car can travel $62^1/2$ km on 5 litres of petrol. A larger model can travel only $4/5$ of this distance. How far should the larger model travel using $6^1/2$ litres of petrol?

5   A county council estimates that by raising the rates by 1p in every £1.00 they would have an extra £1 230 000.00. They propose to spend this on education, police, highways, social services, and miscellaneous projects in the ratio of 76:10:11:12:14. How much would be spent on education?

6   Sheila has two lengths of wire each 88 cm long. She forms a square with one piece and a circle with the other. Which shape encloses the bigger area, and by how many cm²?

7   On the shortest day of the year the sunrise in Bristol was 0813 h and sunset 1603 h. On the longest day sunrise was at 0352 h and sunset 2031 h. How much longer in hours and minutes is the longest day than the shortest day?

8   The cost of sending goods by carrier is advertised at £4.40 per tonne. The load I sent weighed 875 kg. How much did I pay for carriage?

9   Two jars of coffee and three jars of Bovril cost £3.82. At the same prices, three jars of coffee and six jars of Bovril cost £6.75. What is the cost of one jar of Bovril?

10 To cover a kitchen worktop 2.8 m long and 75 cm wide, Lisa bought new Formica at £6.80 per m². A tin of adhesive cost her £2.35. What was the total cost of covering the worktop?

Set 52

1   A new hall door had space for four panes of glass each 55 cm by 40 cm. The patterned glass for the door cost £6.00 per m², and the beading and putty totalled £3.95 for the whole door. What was the total cost of glass and other materials?

2   A grain storage bin measuring 2.5 m by 3.7 m by 1.9 m is only ⅗ full. How many m³ are needed to fill it?

3   Mr Brown was laying the first row of bricks for a new garden wall, 49.13 m long. Each brick measured 22.4 cm and he allowed 1 cm for cement between bricks. How many bricks did he use for that first row?

4   From a piece of copper, Hassan cut ⁴⁄₃₅, then decided he needed only ²¹⁄₂₈ of that for a bracelet he was making. What fraction of the copper did he not use?

5   A glass jar 18 cm tall is half full of acid. The radius of the jar is 4.2 cm. How many cm³ of acid does the jar contain?

6   In a group surgery, four doctors share 9 720 patients. The senior partner as ⅓, the next two have ⅘ of the remainder between them, and the junior partner takes what is left. How many patients does the junior partner have?

7   Houses on an estate which were priced at £32 500.00 in 1977 had risen by 12% by the end of 1979 and by a further 14% by the end of 1981. What was the price of a house on that estate at the end of 1981?

8   Find the total cost of the ingredients of a cake made from 250 g of sugar at 32p per kg, 250 g of butter at 82p per ½ kg, 4 eggs at 72p a dozen, 350 g of flour at 60p per kg, and a carton of cream at 41p.

9   A Boeing jet airliner flew from London to the eastern USA, a distance of 4 225 at an average groundspeed of 650 km/h. It left London at 1935 h. Allowing for 5 hours difference between London and the USA, at what time did it arrive?

10   What is the shortest length of timber that can be cut exactly into either 12 cm, 18 cm, or 30 cm pieces?

Set 53

1   A carpet manufacturer owns ⅗ of his business and decides to sell ⅓ of his share. He then gives ¼ of the remainder to his daughter. What fraction of the business does he still own?

2   The planet Mercury is said to be 56 million km from the sun. The speed of light is 298 000 kilometres per second. To the nearest second, how long does it take a ray of light to reach Mercury from the sun?

3   A mechanical toy has two brass gear wheels, one twice the diameter of the other. The diameter of the larger one is 5.6 cm. What area of brass was needed when cutting out the two wheels?

4   A training run by a cyclist took her from Huddersfield to Hyde at an average speed of 12 km/h and back to Huddersfield at 18 km/h. The whole journey took her eight hours. How far is it from Huddersfield to Hyde?

5   The journey from Liverpool to Preston is quicker on the M6 than on the A59 although the distance is greater. How much farther is it if, on a map with the scale 1 cm to 3.5 km, the motorway measures 18.2 cm and the A59 measures 13.5 cm?

6   Mr White had 7⅘ litres of paraffin. His heaters used 3/13 of this per hour. For how many hours and minutes was he able to use the heaters?

7   An international appeal in a newspaper said that it would cost just 75p to feed a child for one day in an African country. How much would it cost to feed five children for a year?

8   The first King George came to the throne in 1714 and the fourth King George died in 1831. The number of years that each reigned was in the ratio of 3:33:60:11. For how many years did each king reign?

9   A metal drum contains 4 928 cm³ of water which is 32 cm deep. What is the diameter of the drum?

10   Barry had a rectangular piece of glass from which he had to cut three
triangular panes for church windows. The base of each of the first two
windows was 55 cm and their height was 42 cm. The third window was
twice the area of each of the other two. Barry had no glass left over. What
area of glass had he at first?

Set 54

1   Three plots of building land adjoining each other measured 225 m by 106 m,
159 m by 97 m, and 128 m by 159 m. It was decided to divide them into three
equal-sized plots to be sold at the rate of £5 000.00 per hectare. How much
did each plot cost?

2   When cycling from his home to Brighton at an average speed of 15 km/h,
Don covered the distance in 1.75 h. On the return journey, with a following
wind, he averaged 25 km/h. How long did he take on the return journey?

3   A steel rod 215 cm long weighs 6.88 kg. What would be the weight of a rod
with the same diameter, but only 116.1 cm long?

4   A glazier received an order for glass for framing 80 pictures, each 32 cm by
45 cm. The glass cost £2.50 per m². What was the total cost of the glass?

5   16 is a square number because 4 × 4 = 16. Which of the following is a
square number—2 000, 640, 400?

6   When trying to work out the date on a very old church document which read
MDLXXVII, a student mistook the L for a C and the V for an I and recorded
the date as 1623. By how many years was he wrong?

7   One clothing shop allows 2½% discount for cash, while a second shop
allows 3½%. To settle a bill at the first shop, Julian paid £17.55 in cash, but if
he had settled a bill for the same amount at the second shop he would have
had to pay only £17.37 in cash. What was the amount of the original bill?

8   ³/₃₂ of the 960 cars coming off the assembly line needed extra work to be
done on them, and out of these, ²/₉ needed paint re-touching. How many
cars needed paint re-touching?

9   One television set and two transistor radios can be bought for a total of
£348.00. One television set and three transistors at the same price would
cost £373.00. How much is a television set?

10   A 7½-litre water container is ³/₁₀ full. How many litres of water are needed to
fill the container and what is the weight of water in the full container?

Set 55

1 The radius of the earth at the equator is recorded as 6 378.18 km. What is the distance around the earth in km?

2 A length of wood measuring 2.55 m is cut into two pieces so that one piece is four times as long as the other. What is the length of each piece of wood?

3 In the Shell Atlas of Great Britain, Penzance, Redruth, and Bodmin are all on the A30. The map distance from Penzance to Bodmin is 22.4 cm and from Penzance to Redruth is 7.9 cm. The scale of the map is 1 cm to 3 km. What is the actual distance from Bodmin to Redruth?

4 Caroline invested £5 650.00 in her bank at 9% annual interest, but later found that she could have invested it at 11½%. How much would she have lost by the end of one year by investing at the lower rate?

5 How many marmalade jars 10 cm high with a diameter of 12.6 cm could be filled from a container holding 6 237 cm³ of marmalade?

6 Tea trolleys cost a store £28.50. It sold five to customers who paid in 9 instalments of £3.95 each, and it sold off the other five for cash at a 20% profit on each. What was the total profit made on the ten trolleys?

7 An author wrote a book containing 175 pages. Each page had an average of 19 lines, with an average of 8 words per line. It took 40 days to write the book. What was the average number of words written per day?

8 A garden pool measuring 2.8 m by 1.9 m by 0.4 m was absolutely full of water. A concrete block measuring 30 cm by 2 cm by 12 cm was put in the pool as a stand for an ornament. This caused some water to overflow. How many m³ of water remained in the pool?

9 In his will, a man left ⅞ of his money to his wife, ¾ of the remainder to his son, and the rest to charity. He left £3 600.00 altogether. How much was given to charity?

10 A painter had to paint 25 garage doors on a new estate. Each door was 2.5 m high and 3.4 m wide and each needed two coats of paint. One litre of paint should cover 17 m² and a five-litre can costs £13.25. What was the cost of paint for all the garage doors?

Set 56

1 A 255⅗ t load of fertiliser came by rail to be shared by four farmers. The first took ⅓, the second took ½, the third took 1/18, and the fourth took the rest. What weight in tonnes did the fourth farmer take?

2 When re-turfing his lawn, my neighbour bought a load of turfs each 900 cm². His lawn is 12 m long and 9 m wide, and the turfs cost £30.00 per 100. How many turfs did he need and what did they cost him?

3 A store paid £127.80 for a special line in music centres and sold them at a 30% profit. It sold 6 centres in one week. How much money did it take?

4 From the largest number you can make with the digits 4,7,5,3,9, take the smallest number you can make with the same digits and divide the result by 36.

5 I walked a distance of 7.875 km across the Pentland Hills at an average speed of 3.5 km/h. How long did I take for the walk, and how much less time would it take to cycle the same distance at an average speed of 10.5 km/h?

6 Water from a ¾ litre measure was poured into a bowl. When the bowl was full, 378 ml of water remained in the measure. What is the volume of the bowl in cm³?

7 High water at Greenock on 8th January was at 3.39 a.m. and again 11 h 39 min later. A fisherman wanted to catch the high tide with his boat, but he was delayed until 4.30 p.m. Was he early or late? By how many minutes?

8 A settee was sold to a customer who paid £20.00 down and 15 monthly instalments of £18.00 each. By paying cash she could have had a 20% discount. How much would she have saved?

9 It is now possible to travel from Exeter to Carlisle on a motorway for a distance of 470 km. What measurement would this be on a map with a scale of 1 cm to 25 km?

10 Some strips of gold braid, each 12.8 cm long were cut from a length measuring 6.4 m. The total cost was £52.50. What was the cost of each strip?

Set 57

1 Out of a population of 21 030 in a town, 14 721 were over 21 years old. What percentage was under 21?

2 A water tank measures 0.86 m by 0.38 m by 0.22 m. How many litres of water can it hold?

3 Two brothers earned £16 126.00 between them in 1980, but one earned £1 250.00 more than the other. After deducting business expenses, they each had to pay only ⅛ of their income in tax. How much tax did each pay?

4 600 crates with sides measuring 1 m were loaded onto a ship. The ship's hold was 17 m long, 10 m wide and 3 m deep. When it was full, the remaining crates were put on the deck. How many were put on the deck?

5 A customer wanting to buy a carpet costing £182.50 is allowed to pay £30.00 down and the rest in 12 equal instalments of £13.75. How much extra does she have to pay by this method?

6 Two rubbers and three pencils cost 52p whilst three rubbers and four pencils at similar prices cost 72p. What is the cost of each rubber and pencil?

7   A packer has five identical parcels, each 1.5 m long, 50 cm wide, and 10 cm high. He covers them with wrapping paper and allows 1 m² for overlapping on each parcel. The paper costs 37p per m². What is the cost of wrapping paper for the five parcels?

8   A circular concrete path around a garden pond has a circumference of 13.2 m. The circumference of the pond is exactly half of this. What is the width of the path?

9   How many times can 37 be subtracted from ⅛ of 19 240?

10 The approximate lengths of four great rivers is—Nile 6 700 km, Niger 4 200 km, St Lawrence 1 900 km, and Indus 1 700 km. Express these lengths as a ratio.

Set 58

1   A tank measuring 2 m by 1.5 m contained fertilizer pellets to a depth of 1.4 m. What was the value of the fertiliser at £26.00 per m³?

2   Postcards are cut from sheets of card measuring 180 cm by 90 cm and each sheet costs £1.16. What will be the cost of card for 1 350 postcards each measuring 150 mm by 100 mm?

3   A shopkeeper had a sack of potatoes. She sold ⁷⁄₂₅ on Friday and ⁵⁄₁₄ of the remainder on Saturday. What fraction of the whole sack did she sell on Saturday?

4   A fish tank which is 45 cm long and 38.5 cm wide is filled to the brim and it holds 55 440 litres. How deep is the tank?

5   Mrs Baker bought three 50p telephone stamps each week for eleven weeks. When her bill came in it was for £39.73. How much more should she have saved?

6   Last year a father earned three times as much as his son, and between them they earned £20 464.00. Father paid 12% of his salary in mortgage for his house. How much had he left?

7   Two maps of the same area have different scales. The first is to the scale 1 : 25 000, and the second is 1 : 50 000. On the first map the distance between two villages is 1.7 cm. What would be the distance between the same two villages on the second map?

8   ⅞ of an oil tank can be filled in 18⅔ minutes. At this rate, how long would it take to fill the tank to the top?

9   How much greater than 8.698 is ⁹⁄₁₅ of 62.25?

10 In a road walk, the winner covered the distance of 29.25 km at an average speed of 4.5 km/h. For how long was he walking, and how much less time would he have taken if he had averaged 5 km/h?

Set 59

1 During the summer season at a seaside town, an appeal was made for the lifeboat. The collection took the form of a line of 10p coins along the sea front. The line was 52.25 m long. A 10p coin has a diameter of 2.75 cm. How much money was collected?

2 An aircraft can fly at an airspeed of 650 km/h. How long would it take to fly a distance of 2 730 km, and how much longer would it take to fly the same distance if it had a head wind of 50 km/h?

3 A rectangular piece of land $^{24}/_{39}$ km by $^{13}/_{32}$ km in a market garden was divided into eight equal-sized plots for eight different crops. What fraction of a km² was occupied by each plot?

4 A bookshop opened for 5 days in Bank Holiday week. It sold an average of 358 books per day. On the first 4 days it sold 623, 393, 149, and 407 books. How many were sold on the fifth day?

5 A cycle which cost £75.50 last year was increased in price this year by 8p in the £. What is the price of the cycle this year?

6 In a canteen kitchen a circular cake tin has a radius of 12.6 cm and a depth of 15 cm. The cook filled only $^1/_3$ of it with cake mixture. What volume of the mixture was in the tin?

7 What number added to itself 139 times is equal to 25% of 30 024?

8 A long-distance walker had covered $^5/_7$ of the course as he reached 20 000 m. What was the distance in km over which the race was being walked?

9 A tank measuring 40 cm by 40 cm contained diluted acid to a depth of 20 cm. A piece of iron 20 cm by 10 cm by 10 cm was lowered into it for cleaning. By how much was the level of acid raised in the tank?

10 A supermarket had a stock of 1 405 kg of mince in its freezer at the beginning of Christmas week. By Christmas Eve it had sold 77% of this at £1.68 per kg. The rest was offered at £1.40 per kg but only 50% of it was sold. How much money altogether was taken for the mince?

Set 60

1 A sheet of glass $1^5/_{12}$ m long and $1^5/_{34}$ m wide is used for six equal-sized window panes. What is the area of each pane?

2 Out of a consignment of 750 plates sent to a china shop, 45 were broken. What percentage was unbroken?

3 9^3 plus another number equals 793. What is the other number?

4   A grocer bought a cask of vinegar containing 22⁴/₅ litres for £14.00. He bottled the vinegar in 200 ml bottles and sold each for 19p. If he paid 2p for each bottle, what profit did he make?

5   A lean-to greenhouse has six square panes of glass, each 55 cm square for the front, six for the roof, and two at each end. One pane is cut diagonally, with half for each end. How many m² of glass are needed for the whole greenhouse?

6   A blocked drainpipe 2 m long, with a radius of 3.5 cm is ⁷/₁₀ full of water. How many litres does it contain?

7   A water tank contains 3 724 litres when full. It is 2.45 m long by 1.6 m wide. How deep is it?

8   At the beginning of this year at a city college there was a total of 1 976 students, with twelve times as many men as women. By the end of the year, ¹/₈ of the women and five times that number of men had dropped out. How many men were left?

9   The ferry from Kyle of Lochalsh to Skye is £3.00 cheaper than from Mallaig to Skye. A car driver at Spean Bridge measured the distance to both ports on a map with a scale of 1 cm to 6 km. To Lochalsh was 13.8 cm and to Mallaig was 10.1 cm. Which would be the cheaper route and by how much if he allowed 10p for every km he motored?

10   A winter cruise ship left Southampton at 0830 h on 22nd January and arrived at Colombo, Sri Lanka, 21 days, 4 hours, and 15 minutes later. Allowing for Sri Lankan time being 5½ hours ahead of British time, what was the date and time of the ship's arrival at Colombo?

Set 61

1   It has been calculated that an average of 13½% of eggs that are laid will be cracked or broken by the time they reach the shops. The manager of a poultry farm used this information to estimate how many eggs, out of the 950 dozen that were laid in one week, would be fit for sale by the time they were in the shops. What was his estimate?

2   A rubber sheet 10 cm long and 8 cm wide is stretched so that it now measures 15 cm by 12 cm. One boy said that its area must now be 1½ times as big. By how many cm² was he wrong?

3   The journey from Perth to Stirling takes 84 minutes on a moped at an average speed of 30 km/h. Julie wanted to reduce the time taken by 24 minutes when she went on her moped. At what speed did she need to travel?

4   What number is a quarter of the way between 54² and 24³?

5 At the local show a canvas screen 1.8 m high was put up to enclose the refreshment bar. It had three sides—two were 7.5 m long and one was 8 m long. The canvas cost £3.75 per m². What was the total cost of the screen?

6 A factory supervisor earns £8 955.00 per annum and a labourer earns £5 607.00. They are both due for an increase of 8%. How much greater will the supervisor's increase be than the labourer's?

7 Mark worked out that in one period of 24 h he slept, worked, ate, watched television, and played with his friends in the ratio of 30:25:8:5:4. How much time was left for other activities?

8 A newspaper reporter wanted space for his story in the local paper. The editor said that he could have 105 lines out of one column. In that paper there are seven columns to a page and 180 lines to a column. What fraction of a page was the reporter able to use?

9 In a clockwork toy, two gear wheels engage each other. The smaller of the two has to turn eight times to every single turn of the larger wheel. The circumference of the larger wheel is 704 mm. What is the radius of the smaller wheel?

10 The manager of a furniture shop bought 15 settees for a total of £2 175.00. She sold them for £188.50 each. What percentage profit did she make on each?

Set 62

1 In a high school, exactly half of the children were girls. Of the girls, $^{10}/_{19}$ were aged 14 years or over, and out of these, $^{38}/_{45}$ were studying for GCE or CSE examinations. What fraction of the whole school was made up of girls taking these examinations?

2 Shirley wanted to go from Hull to Carlisle. On a road map with a scale of 1 cm to 10 km she measured the distance as 8.5 cm, but she thought she had measured it on a map with a scale of 1 cm to 5 km. She estimated that she could do the journey in 8½ hours if she averaged 50 km/h, but how long, in fact, would the journey take at that speed?

3 An employer of seven workers paid them an average of £303.00 each per month. The first three received £450.00, £389.00, and £342.00, and the remaining four shared the rest equally among themselves. How much did each receive?

4 Divide the product of 5, 7, 13, and 15 by $^5/_8$ of their sum.

5 The area of the bottom of a plastic cube is 156.25 cm² and it is 12.5 cm tall. How many litres of water will the cube hold when full? Answer to the nearest litre.

6   An exchange bureau accepted £153.50 to change into French francs in 1980 at 10.80 francs to the £. How many francs were received for the English money?

7   The area of a square garden is 6 400 m². What would be the cost of putting a fence completely around it at £3.75 per metre length?

8   A can holding powdered milk has a diameter of 98 mm and a height of 16 cm. ⅛ of the powdered milk has been used. What volume is left in the can?

9   A measure holds ⅝ litres. How many times can I fill the measure from a container holding 16⅛ litres and how much water will be left in the container?

10 On a scale drawing, an architect draws a line 32 mm long to represent a distance of 25.6 m. What scale is he using? (What does 1 cm represent?)

Set 63

1   A lorry driver drove for two hours at an average speed of 70 km/h before going on to the motorway where he increased his speed to 80 km/h. He reached his destination 1⅗ hours later. How far had he travelled?

2   What is one tenth of the difference between 25^2 and 25^3?

3   On a map with a scale of 5 cm to 1 km the distance between two villages is 16.5 cm. What would the distance be on a map with a scale of 8 cm to 1 km?

4   Two sacks of rice, one holding 32½ kg and the other 38¼ kg are packed into boxes each holding 1⅔ kg. How many boxes could be packed and how much rice would be left over?

5   Emma bought a piece of foam rubber for a settee seat 2 m long, 75 cm wide, and 10 cm thick. She covered this with patterned material costing £4.40 per m², and she allowed 5 m² for overlap and sewing. What did the material cost her?

6   British Columbia time is 8 hours behind London, Manitoba is 6 hours, and Newfoundland 3½ hours behind. What will be the time in each of these three provinces when it is 1400 h in London?

7   The County Council was discussing planning permission for a piece of land 13/15 km long, 15/26 km wide. ⅘ of this was to be given to housing and the rest to a small factory. What fraction of a km² was given to the factory?

8   When railway fares were increased from 2½p per km to 3¼p per km, what was the new cost of a journey which cost £7.50 before the increase?

9   A plastic bottle when empty weighs 28.7 g. When half full of water it weighs 96.55 g. What weight of water will the bottle hold?

10   A fireworks factory uses the following ingredients when making
     gunpowder—sulphur 15%, charcoal 15%, and nitre 70%. What quantity of
     sulphur would be needed when making 25k g of gunpowder?

Set 64

1   This year after tax a man had 72% of his income to spend. He paid £2 352.00
    in tax. What was his income?

2   A cart wheel has a diameter of 602 mm. What distance, in metres, has it
    travelled when it has turned 100 times?

3   How many times will 33^3 go into 66^3?

4   The Continental Express from Paris travelled at an average speed of
    92.25 km/h for 7.85 h. What distance had it covered in that time? If it left
    Paris at 0630 h, what was the time 7.85 h later?

5   In one day at a seaside resort a survey was made of the people who used
    deckchairs. Children used only $^1/_{16}$, and of the remainder, women used $^7/_{25}$
    and the rest were used by men or they remained empty. What fraction of
    the total number of deckchairs was used by women?

6   By paying cash I could buy a wall cupboard for £153.00. If I pay in 18
    monthly instalments it will cost me £6.30 extra. How much is each monthly
    instalment?

7   A tanker is $^7/_8$ full when it holds 6 846 tonnes of oil. What weight does it hold
    when it is $^2/_3$ full?

8   A water drainpipe has an outside diameter of 3.5 m and is 16 m long. What
    volume of soil will have been removed in order to bury it in the ground?

9   A drinking trough for farm animals contains 600 litres of water. The area of
    the bottom of the trough is 3m². How deep is the water in the trough?

10 A length of insulated wire is cut down by 10% and now measures 8.1 m.
    What would be its length if it had been cut down by 20%?

Set 65

1   A map to the scale of 1 : 25 000 shows a stretch of canal between two locks
    as being 4.5 cm. What is its actual length and how many cm would it
    measure on a map with a scale of 1 : 50 000?

2   Walter is half as tall as his sister Jenny who is $^4/_5$ as tall as her elder brother
    Michael. Michael is $^8/_9$ as tall as his father who measures 1.8 m. How tall is
    Walter?

3   One 50p coin weighs 13.5 g. What is the total value of a pile of 50p coins weighing 675 g?

4   From a cask of vinegar containing 3.1 litres, eight 200 ml bottles are filled. How many 250 ml bottles can now be filled from what is left in the cask?

5   A staircase is 60 cm wide. It has 15 stairs and each one measures 15 cm vertically, 23 cm horizontally. Mr Copland buys carpet to cover the stairs completely, starting 50 cm from the bottom stair and allowing an extra 23 cm from the edge of the top stair onto the landing. How much does the stair carpet cost him at £8.00 per m²?

6   Caroline had been saving coins in her piggy bank for over a year. She now wanted to buy a secondhand cycle costing £45.25. In her bank she had twenty 50p, nine 20p, twenty-one 10p, one hundred and twenty-six 5p, two hundred and forty-seven 2p, and five hundred and seven 1p coins. How much more money did she need in order to buy the cycle?

7   A square field with an area of one hectare is surrounded by a fence consisting of three rows of barbed wire. What length of wire was used, allowing an extra 25 m for joins?

8   A father gave some money to his three sons. The average amount that they had was £6.00, but the eldest took three times as much as the youngest and then shared his amount equally with the middle son. How much did each have?

9   A pile of books is 97⅕ cm high. Each book is 3⅗ cm thick. How many books are there in the pile?

10 During the five months from March to July, a town in India had a total of 1 950 mm of rain, in the ratio of 3 : 7 : 15 : 19 : 21. How much more rain did the wettest month have than the driest?

Set 66

1   A home aquarium 31.5 cm long, 21.6 cm wide contains water to a depth of 13 cm. A can 15.5 cm high with a diameter of 11.2 cm, is completely filled from the aquarium. How many cm³ of water are left?

2   Two aircraft are flying in opposite directions on an east–west route over a distance of 2 700 km. There is an easterly wind of 20 km/h. Both aircraft have an airspeed (speed through the air) of 520 km/h. How much more time will be taken for the journey by the one going east than the one going west?

3   The inside diameter of an iron drainpipe is 140 mm and the outside diameter is 154 m. What volume of iron is contained in each metre length of pipe?

4   Darning needles in bundles of five weigh 1⅞ g. How many bundles of needles can be made from a total of 66⅜ g and how many needles would be left over?

5   A reel of tape 5 m long is cut into 80 pieces which are put into two piles of 40 pieces each. The pieces in the first pile are half as long again as the pieces in the second pile. What are the two lengths into which the tape is cut?

6   The earth turns through 360° of arc in 24 h. How long does it take to turn through 1°?

7   Divide the product of the first five prime numbers into the product of the next five.

8   A motor scooter covers a distance of 149⅝ km in 4½ h. What is its average speed in km/h?

9   Gillian invested £6 600.00 for one year at the following rate—⅓ at 9½%, ⅓ at 11%, and ⅓ at 14%. How much interest had her money gained by the end of the year?

10   The customs duty on transistor radios is 33⅓%. How much did a dealer have to pay on a consignment of radios valued at £174.00?

Set 67

1   Eleven pieces of copper wire each 25 mm long are cut from a length of 50 cm. How many pieces each 15 cm long can be cut from the remainder?

2   In a marathon cycle race of 420 km, the winner travelled at an average speed of 16 km/h and the last cyclist averaged 14 km/h. How much longer did the last cyclist take than the winner?

3   A pile of cork tiles, each ⅚ cm thick, is 121⅔ cm high. How many tiles are in the pile?

4   A sports store bought a gross of tennis balls for £90.72. They were sold to make a total profit of £17.28. What was the cost of each ball to the customer?

5   The outside circumference of a tyre is 1.76 m and the inside circumference is 132 cm. What is the thickness of the tyre?

6   A farmer had two rectangular fields. One measured 680 m by 560 m and the other 710 m by 645 m. 5 hectares were fenced off for experimental crops. How many hectares were left for other crops?

7   Father weighs 96 kg while mother weighs ¾ of this. The older son weighs ⁷⁄₉ of mother's weight and the younger son weighs ⅝ of his older brother's weight. What is the weight of the younger son?

8   John wanted to make a scale drawing of a building 105 m long by 79 m wide. He wanted to use A4 paper, which is 296 mm by 210 mm. He needed to choose a scale that would allow him to make the biggest possible drawing. He tried 1 cm to 1 m, 1 cm to 2 m, and so on up to 1 cm to 6 m. Which should he have chosen?

9 In a walking race the first four to finish completed the course in 1 h 43 min, 1 h 56 min, 2 h 05 min, and 2 h 14 min. Only five altogether completed the course, and the average time was 2 h 05 min. How long did the fifth walker take?

10 Gary invested £1 000.00 in a building society to gain 10% interest each year. At the end of the first year he left the interest in the building society so that it, too, gained 10% interest, and so on for five years. By how much had his £1 000.00 grown by the end of the five years?

Set 68

1 A chemist had 48 litres of liquid shampoo which he put into three sizes of bottles containing 750 ml, 500 ml, and 250 ml. He filled an equal number of each size. How many bottles altogether did he fill?

2 A headteacher has two rings, one inside the other, marked on the school playground. The first has a circumference of 308 cm and there is space of 35 cm between the two rings. What is the circumference of the outer ring?

3 When planning a garden, a gardener had 46 $3/7$ m² in which to plant roses. She decided to allow 1$6/7$ m² for each rose bush. How many rose bushes did she buy?

4 The total amount contained in two bags of money is £15.40 and their difference in value is £4.80. How much money is in each bag?

5 A lawn has an area of 661$1/2$ m² and is 24$1/2$ m wide. It costs 55p per metre to put a protective surround around the lawn. What will the cost of this be if the delivery charge is £1.20?

6 A 10p coin has a diameter of 2.8 cm and a thickness of 2 mm. What volume of metal, in cm³, would be needed to mint fifty of these coins? ($\pi = 3^1/7$.)

7 A toy maker was making 125 wooden cubes with sides of 3 cm for use as toy bricks. The wood he bought was already planed to measure 3 cm by 3 cm, but what length did he need, and what was the total volume, in cm³, of wood used for the bricks?

8 A glass cylinder is 30 cm tall with a diameter of 15.4 cm. How many times could this be completely filled from a tank measuring 40 cm long by 35 cm wide, which contained water 20 cm deep? How much water would be left in the tank?

9 A shop had fruit juice for sale in two different sized cans. The smaller size was 11 cm tall, diameter 63 mm, and the larger was 14 cm tall, diameter 77 mm. The smaller tin cost 54p and the larger £1.08. Which was the better buy; two of the smaller or one of the larger tins, and how much more fruit juice would there be?

10   Litres of emulsion paint are packed into crates. Each full crate costs the hardware store £80.88. It is sold for £4.25 per litre, giving a profit of 88p per litre. How many litres are in a crate?

Set 69

1   A car travelling at an average speed of 72 km/h covers a distance of 460.8 km. A second car takes 48 minutes longer for the same distance. What is the average speed of the second car?

2   A city roundabout is 23.8 m in diameter and has on it four circular flower beds of radius 3.5 m each. The rest of the roundabout is concreted to a depth of 5 cm. What volume of concrete was used for this?

3   A greengrocer sold $5/8$ of his stock of potatoes on Monday, $16/21$ of the remainder on Tuesday, and $1/5$ of what was left on Wednesday. What fraction of his original stock was sold on Wedneday?

4   The sides of a hollow cube made of plywood measure 2.15 cm. What area of plywood is needed to make five of these cubes?

5   Lino tiles 24 cm square are sold in boxes of 22 for £4.85. My bathroom is 2.64 m long, 1.92 m wide. Allowing a space of 1.2 m by 0.72 m for the bath, etc, if I am allowed to buy only full boxes of tiles, how much will I have to pay for tiles for the bathroom and how many will be left over?

6   Three aircraft, A, B, and C, fly for a distance 1 120 km. A flies at 560 km/h, B flies at $5/8$ the speed of A, and C flies at $5/7$ the speed of B. How many hours and minutes will each take to cover the distance?

7   A distance of 100 km is represented on a map by 5 cm. What is the scale of the map, and how many cm on the map would represent a stretch of motorway 430 km long?

8   Kate left her home station at 1645 h on Thursday and arrived at Harwich 2 h 16 min later. She waited 1 h 20 min for her ferry to sail, and it took 6 h 59 min to reach the Hook of Holland. She waited 3 h 42 min for her train, which then took 1 h 41 min to Brussels. At what time and day did she arrive in Brussels?

9   Mr McHugh spent the same amount on food in January as in February. He calculated that the daily average was £5.60 in January. What was it in February (not a leap year)?

10   What is the ratio of 120^2 to 160^2?

Set 70

1   A triangular pyramid is made up of 4 equilateral triangles each with a base of 10 cm and a perpendicular height of 8.6 cm. Each of the 21 children in a class makes a card pyramid of this size, using Sellotape to join the edges. What area of card is used altogether?

2   In the centre of a square lawn with sides of 5 m is a circular pond with a circumference of 6.6 m. It has a concrete surround, 0.7 m wide. What area of grass is around the concrete?

3   Sauce which costs the manufacturer £1.20 per litre to make is sold in the shops at 84p per 250 ml bottle. The manufacturer makes 15% profit when selling to the shopkeeper. What profit per litre does the shopkeeper make?

4   What is the highest square number below 2 100?

5   Sonya invests some money in a bank at 10% interest per year. At the end of the first year she decides to leave the original sum and the interest for another year. At the end of that time she has a total of £1 210.00. How much did she invest at first?

6   A wooden tube has an inside diameter of 6.3 cm and an outside diameter of 7.7 cm and it is 16.6 cm long. What volume of wood has gone into the making of the tube?

7   On a map with a scale of 1 cm to 50 km, the distance between two towns is represented by 12.4 cm. On a second map the same distance is represented by 31 cm. What is the scale of the second map?

8   A man left some money in his will to his four daughters. The youngest had £160.00, which was $\frac{1}{12}$ of the total. The next daughter had $\frac{1}{8}$, the next $\frac{1}{4}$, and the eldest daughter had the rest. How much did the eldest daughter have?

9   A church wall was bulging outwards. A stone and cement buttress 20 cm wide was made. It started 2.16 m from the base of the wall and went diagonally up to a point 3.64 m from the ground. What volume of stone and cement went into the buttress?

10   What number squared is equal to $\frac{1}{25}$ of 75^2?